MANAGEMENT INFORMATION SYSTEM

NEHA GHAI
B.Tech.
Dronacharya College of Engineering, Gurgaon

S.K. KATARIA & SONS®
Publisher of Engineering Books
4885/109, Prakash Mahal, Dr. Subhash Bhargav Lane,
Opposite Delhi Medical Association, Daryaganj, New Delhi–110002 (INDIA)
Phone: +91-11-43551243, *Mobile:* +91-9871775858, 7678425209
e-mail: katariabooks@yahoo.com, katariabook@gmail.com
Website: www.skkatariaandsons.com
(e-books,subscriptions and bulk buying also available on our website)

Published by:

S.K. Kataria & Sons®

4885/109, Prakash Mahal, Dr. Subhash Bhargav Lane,
Opposite Delhi Medical Association, Daryaganj, New Delhi-110002 (INDIA)
Phone: +91-11-43551243; *Mobile:* +91-9871775858, 7678425209
e-mail: katariabook@gmail.com, katariabooks@yahoo.com
Website: www.skkatariaandsons.com

First Edition: August, 2012; Reprint: 2026

Price: ₹ 425/-

Laser Typesetting by :
PLUS COMPUTERS, Shahdara, Delhi-110093.

Printed at :
New A.S. Printing Press, Delhi–110051

Preface

Management information system has become a crucial factor in the working of different business organizations.

This book provides the basics to students about management information system and its use in different organizations.

It covers the modules related to management information system, information systems, information technology and advanced management information systems in a quite comprehensive way.

Keeping in mind the needs of the students, the book covers all the topics of the syllabus in a simple manner.

For systematic and organized study, the book is divided into four sections and consists of fourteen chapters in all. At the end of each chapter; summary, solved question answers and review questions are provided. The book also provides with unsolved and solved previous year university question papers.

Suggestions, if any, are always welcomed from the readers.

Neha Ghai

Acknowledgement

Nobody can do anything without the grace of God. I first thank "LORD KRISHNA" who blessed me to achieve my ambition by writing this book.

I am grateful to my grandmother, parents and brother for supporting and encouraging me in accomplishing my goal. I would also like to thank my friends who supported me in each step.

I convey my thanks to the publisher "M/s S.K. Kataria & Sons", who gave me the golden opportunity to write the book.

Neha Ghai

Contents

Section-A

Inside This Section :

Chapter 1

Introduction to Information Systems

"Not everything that can be counted counts and not everything that counts can be counted."

Inside this Chapter

1.1. INTRODUCTION TO AN INFORMATION SYSTEM

Definition : An Information System is an organized combination of people, hardware, software, resources and communication networks. Information system is also a scientific study that covers strategic, managerial and operational activities. These three activities basically deal with the following operations :

- Gathering of Information
- Processing of Information
- Storage of Information

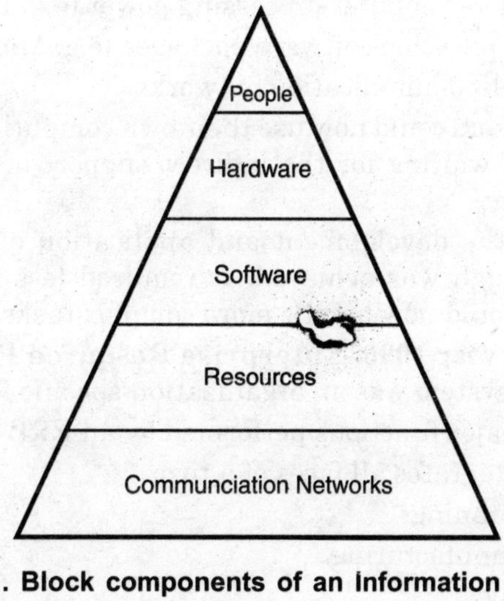

Fig. 1.1. Block components of an Information System

In an Information system, there is basically a coordination between people and technology to achieve a common goal. It also includes the organization's manual and automated processes.

Information system provides assistance to management. This is done by providing a feedback on the firm's performance.

> ✍ **Note:** Feedback is the output of the system that is transformed back into the input so that the system's operation can be controlled.

1.2. EVOLUTION OF INFORMATION SYSTEMS

In the era of **1960**, the role of Information System was very simple. They were mainly used for electronic data processing's like transaction processing, record keeping and accounting purposes.

Electronic Data Processing : It is the use of computers in performing the following operations :

- Recording of data
- Classification of data
- Manipulation of data
- Summarizing of data

In the **1960's**, another role was added to the use of computers. This role was the processing of data into useful informative reports. Here, the concept of **Management Information System** (MIS) came into existence. This concept focused on developing business applications that provided managerial end users with predefined management reports. These predefined management reports provided managers with the information that is required for performing decision making processes.

But in the end of **1970's,** these pre-defined management reports were not sufficient to meet many of the decision-making needs of the management. The concept of **Decision Support System** came into existence to satisfy the needs that were not fulfilled earlier.

The new role for Information system was to provide managerial end users with interactive support of their decision-making processes.

In the era of **1980's**, microcomputers came into existence. It had a great impact on the organizations. End user computing also came into force by the development of :

- Microcomputer processing power (e.g. Intel's Pentium Microprocessor)
- Application software packages (e.g. Microsoft Office)
- Telecommunication networks

End users could now use their own computing resources to support their job requirements instead of waiting for the indirect support of a centralized corporate information services department.

With the development and application of artificial intelligence techniques, a further breakthrough was achieved. It required less need of human intervention and knowledge. Workers could now handle more complex tasks.

In the year **1990**, **Enterprise Resource Planning (ERP)** systems also came into force. The ERP system was an organization-specific form of strategic information system.

The major functions performed by an ERP system are briefly mentioned below :

- Integrates all faces of a firm
- Planning
- Manufacturing
- Sales

- Resource management
- Customer relations
- Inventory control
- Order tracking
- Financial management
- Human resources
- Marketing

It provides a common interface for all computer-based organizational functions.

1.3. VIEWS OF AN INFORMATION SYSTEM

In 1995, Silver et al. gave two views on an Information system.

The **first view** includes the following components :

- Software
- Hardware
- Data
- People
- Procedures

The **second view** includes the following components :

- People
- Business process
- Information system

The people associated in an Information system are the trained personnel. The major responsibilities of these trained personnel are :

- Planning of activities
- Control and Coordination of activities
- Decision Making
- Decision Support
- Performance of business process

1.4. COMPONENTS OF AN INFORMATION SYSTEM

The main components of an Information system are listed below :

- **Computers :** Nowadays computers are used is in both small and big firms. Large organizations make use of distributed computer systems. The distributed computer system consists of powerful processing servers that are located in various data centers. These also consist of peripheral equipment's like storage disks and input-output devices.
- **Facts :** Fact is either that has really occurred or is actually the case.
- **Instructions :** Instruction is a sequence of bits that instructs the central processing unit to perform a particular operation. It may also contain data that can be used in the operation.
- **Procedures :** A procedure is basically a set of operations that is to be executed in the same manner so that the same result is obtained in all situations.

1.5. TYPES OF INFORMATION SYSTEM

There are various types of Information system. These are listed below :

- Transaction Processing System
- Management Information System
- Decision Support System
- Executive Information System
- Operation Support System
- Knowledge Management System
- Expert Systems

1.5.1. Transaction Processing System

Transaction processing system is basically used in performing the following operations :

- Processing of data that is obtained as results from business transactions.
- Update operational databases.
- Production of business documents
- Process routine transactions in an efficient and accurate manner.

The transaction processing system basically deals with the workers.

The **examples** of transaction processing system are :

- Sales Process
- Inventory Processing
- Accounting Systems.

Types of Transaction Processing System : There are various types of transaction processing system. Each system has its own functions and applications. These are :

I. **Sales/marketing Systems :**
Functions : Sales management, market research, promotion and pricing.
Applications : Sales order and market research.

II. **Manufacturing/ production Systems :**
Functions : Scheduling, purchasing, engineering, shipping and operations.
Applications : Machine order, purchase control and quality control.

III. **Finance/ accounting Systems :**
Functions : Budgeting, billing and cost accounting.
Applications : Accounts receivable and payable.

IV. **Human Resource Systems :**
Functions : Compensation, training and benefits.
Applications : Payroll and employee records.

1.5.2. Management Information System

MIS provides information in the form of pre-specified reports and supports business decision making. It is concerned with the internal sources of Information. The Management Information System deals with the middle managers and operational supervisors. It takes data from the transaction processing system and summarize into a series of management reports.

The **examples** of Management Information system are :

- Sales analysis
- Production

- Performance
- Cost trend reporting systems.

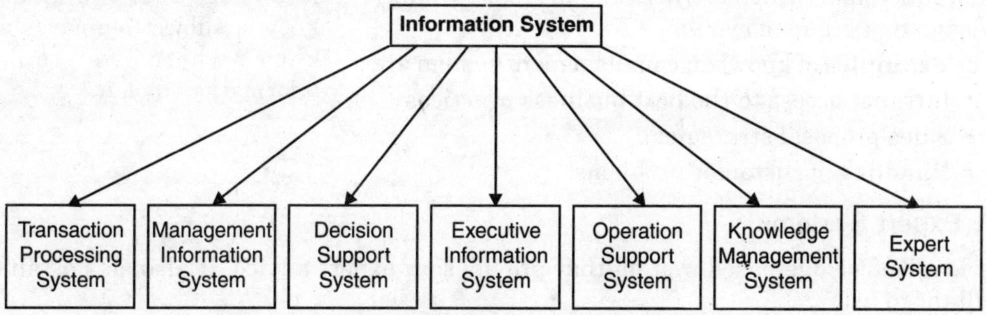

Fig. 1.2. Types of Information system

1.5.3. Decision Support System

Decision support system provides interactive and ad hoc support for the decision-making processes of managers and other business professionals. The decision support system deals with the senior managers. It helps the managers to take decisions in those situations where one is not sure about the possible result of the decision. It contains tools and techniques that help to gather relevant information and analyze the options. It makes use of spreadsheets and databases.

The examples of decision support system are :

- Product pricing
- Profitability
- Forecasting
- Risk analysis systems.

1.5.4. Executive Information System

Executive Information system provides important information from MIS, DSS and other sources according to the information needs of executives. It helps the senior management to take strategic decisions. It gathers, analyzes and summarizes the internal and external information that is used in business processes.

The **examples** of executive information system are :

- Actions of the various competitors
- Economic developments to support strategic planning

1.5.5. Operation Support System

Operation Support System allows IT personnel to administrate operational processes that focusses on the network. It also includes services like :

- Service quality monitoring
- Network performance
- Server performance
- Logical and physical resource management.

1.5.6. Knowledge Management System

Knowledge management system helps the businesses to create and share important information. It is a knowledge-based system that supports the creation and organization of business knowledge within the enterprise. The knowledge is shared with other people in the organization to create commercial opportunities. Intranet is used to share knowledge in a group of system.

> **DO YOU KNOW?**
>
> Knowledge is of two kinds : we know a subject ourselves, or we know where we can find information upon it.

The **examples** of knowledge management system are :

- Intranet access to the best business practices
- Sales proposal strategies
- Handling of customer problems.

1.5.7. Expert Systems

It is a Knowledge-based system that provides an expert advice. It also acts as an expert consultant to users.

The **examples** of expert systems are :

- Credit application advisor
- Process monitor
- Diagnostic maintenance system

1.6. WORKING AREAS AND WORKING METHODOLOGY OF AN INFORMATION SYSTEM

Information system works on the following areas :

- Strategy
- Management
- Development
- Security
- Organization

The instructions and procedures are applied on the facts to create new information and manage the existing information.

It gives more importance to the functionality than design.

1.7. ROLE OF INFORMATION SYSTEMS

The role of a Information system is explained in the following manner :

A. Information Storage and Analysis

Earlier companies used to manage their data manually with the help of registers and hard copy formats but nowadays at the time of publication, various companies manage their data with the help of comprehensive and sophisticated databases. An Information system performs the following operations :

- Storage of Information
- Updation of Information
- Analyzing the Information

After all the operations are performed, the company can then handle the current problems and the problems that may occur in future.

B. Assist With Making Decisions

The management team of an organization makes use of the Information system to perform the following operations :

- Formulate the strategic plans
- Make decisions for the benefits of an organization

The long-term success of a company solely depends upon the adequacy of its strategic plans. The business process makes use of the Information system to evaluate the information that is obtained from all the sources including the external references. This analysis and comparison helps the organizations in analyzing the adequacy and quality of their strategic decisions.

C. Assist With Business Processes

Information system assists with the business processes in developing a large number of value added systems in the company. This can be explained with the help of an example: A company can integrate an Information system with its manufacturing cycle to ensure that the output produced matches with the requirements of the various quality management standards. The benefits of doing so are the following :

- Simplifies business processes
- Removes unnecessary activities
- Only users with applicable rights can perform certain tasks
- Add controls to the employee processes
- Employees can concentrate on high level functions
- Increases accuracy
- Eliminates repetitive tasks
- Better project planning and implementation
- Effective monitoring

D. Automation of Manual Tasks

Information system architecture also helps an organization in automating the manual tasks. Doing so helps the organization to save :

- Time
- Money
- Resources

It also enhances the organizational workflow. There are various types of Information systems that automate the manual processes. These are :

- **Robotic Information systems :** These are used in areas such as health and medical services.
- Logistical Information systems
- Distribution systems

E. Hardware and Software Integration

An organization consists of different computer platforms (hardware and software). The concept of information systems as a scalable platform can combine different hardware and software systems. It helps in the processing, storing and distributing of information. An open architecture information system allows integration at all the levels in an organization.

F. Support of a Multi-Processing Environment

An information system can also support a “real-time” multi-processing environment by using the concept of “time-sharing application.” Time-sharing application allows the prioritizing of applications which are based upon the :

- User-id
- System priority which is assigned to a device.
- System priority which is assigned to a database.
- System priority which is assigned to a system catalog.

All these features are important to an organization that processes transactions while developing and testing program applications. In a multi-processing environment, various departments, divisions or branches can have access to the system at the same time.

1.8. DISADVANTAGES OF INFORMATION SYSTEM

Following are the constraints of an Information System :

- **Technical barriers :** There is no common standard for systems. Each hardware maker makes use of different standards.
- **Individual resistance :** Managers doesn't make complete use of the technology. Managers are either afraid of the technology or there is a lack of knowledge of the existing technology that is in use.
- Information which is rich in meaning is not suitable to be handled by an information system.
- Some information systems are hard to develop and install.

Summary

- An Information System is an organized combination of people, hardware, software, resources and communication networks.
- Information system is a scientific study that covers strategic, managerial and operational activities.
- Management Information System focused on developing business applications that provided managerial end users with predefined management reports.
- The first view of an Information system includes software, hardware, data, people and procedures.
- The second view of an Information system includes people, business process and Information system.
- Transaction processing system, management Information system, decision support system, executive Information system, operation support system, knowledge management system and expert systems are the various types of Information system.
- Computers, facts, procedures and instructions are the components of an Information system.
- Sales, manufacturing, human resource and finance systems are the types of transaction processing system.
- The Management Information System deals with the middle managers and operational supervisors.
- The examples of Management Information system are sales analysis, production, performance and cost trend reporting systems.
- Decision support system provides interactive and ad hoc support for the decision-making processes of managers and other business professionals.
- The decision support system deals with the senior managers.

- The examples of decision support system are product pricing, profitability, forecasting and risk analysis systems.
- Executive information system provides important information from MIS, DSS and other sources according to the information needs of executives.
- Executive information system helps the senior management to take strategic decisions.
- Operation Support System allows IT personnel to administrate the operational process that focusses on the network.
- Knowledge management system helps the businesses to create and share important information.
- Expert system is a knowledge-based system that provides an expert advice. It also acts as an expert consultant to users.
- Information system works on strategy, management, development, security and organization areas.
- Information system assists with the business processes in developing a large number of value added systems in the company.
- Information systems architecture helps an organization in automating the manual tasks.
- An information system can also support a real-time multi-processing environment by using the concept of time-sharing application.
- Strategic, Managerial and operational activities deals with the gathering, processing and storage of information.
- In 1960"s, Information Systems were used for electronic data processing like transaction processing, record keeping and accounting purposes.
- Instruction is a sequence of bits that instructs the central processing unit to perform a particular operation.
- The transaction processing system deals with the workers.
- The management team of an organization uses Information System to formulate the strategic plans and make decisions.

Short Question Answers

1. Define Information System.

Ans. An Information System is an organized combination of people, hardware, software, resources and communication networks. Information system is also a scientific study that covers strategic, managerial and operational activities.

2. What are the components of an Information system?

Ans. People, hardware, software, resources, communication network and resources are the components of an Information system.

3. What are the major functions performed by an ERP?

Ans. The major functions performed by an ERP system are planning, manufacturing, sales, resource management, customer relations, inventory control, order tracking, financial management, human resources and marketing.

4. What is the first view of an Information system?

Ans. The first view of an Information system includes software, hardware, data, people and procedures.

5. What is the second view of an Information system?

Ans. The second view of an Information system includes people, business processes and Information system.

6. What are the responsibilities of the trained personnel of an Information system?

Ans. The major responsibilities of the trained personnel are planning of activities, control and coordination of activities, decision making, decision support and performance of business process.

7. What are the components of an Information system?

Ans. Computers, facts, procedures and instructions are the components of an Information system.

8. What are the types of transaction processing system?

Ans. Sales, manufacturing, human resource and finance systems are the types of transaction processing system.

9. What are the various types of Information system?

Ans. Transaction processing system, management Information system, decision support system, executive Information system, operation support system, knowledge management system and expert systems are the various types of Information system.

10. What is a Management Information system?

Ans. MIS provides information in the form of pre-specified reports and supports business decision making. It is concerned with the internal sources of Information. The Management Information System deals with the middle managers and operational supervisors. It takes data from the transaction processing system and summarize it into a series of management reports.

11. What are the examples of Management Information system?

Ans. The examples of Management Information system are sales analysis, production, performance and cost trend reporting systems.

12. What is a Decision support system?

Ans. Decision support system provides interactive and ad hoc support for the decision-making processes of managers and other business professionals. The decision support system deals with the senior managers.

13. What are the examples of Decision support system?

Ans. The examples of decision support system are product pricing, profitability, forecasting and risk analysis systems.

14. What is an executive Information system?

Ans. Executive information system provides important information from MIS, DSS and other sources according to the information needs of executives. It helps the senior management to take strategic decisions. It gathers, analyses and summarizes the internal and external information that is used in the business processes.

15. What is an operation support system?

Ans. Operation Support System allows IT personnel to administer the operational processes that focusses on the network.

16. What is a knowledge management system?

Ans. Knowledge management system helps the businesses to create and share important information. It is a knowledge-based system that supports the creation and organization of business knowledge within the enterprise. The knowledge is shared with other people in the organization to create commercial opportunities.

17. What is an expert system?

Ans. It is a Knowledge-based system that provides expert advice. It also acts as expert consultants to the users.

18. What are the working areas of an Information system?

Ans. Information system works on strategy, management, development, security and organization areas.

19. How an information system assists with business processes?

Ans. Information system assists with the business processes in developing a large number of value added systems in the company.

20. What is the hardware and software integration of an information system?

Ans. An organization consists of different computer platforms *i.e.,* (hardware and software). The concept of information systems as a scalable platform can merge different hardware and software systems. It helps in the processing, storing and distributing the information.

Review Questions

1. Define an Information system in detail.
2. Explain the evolution of an Information system.
3. What are the disadvantages of an Information system?
4. Explain the types of Information system in detail.
5. Explain the role of Information system in detail.

□□□

Chapter 2

Introduction to Management Information Systems

"I find that a great part of the information I have was acquired by looking up something and finding something else on the way."

Inside this Chapter

2.1. INTRODUCTION TO MANAGEMENT INFORMATION SYSTEM

A management information system is designed by an organization so that it can be used by the managers in performing the following functions :

- Directing
- Planning
- Coordinating
- Communicating
- Decision making

The management information system is used by the top management and consists of various controls. This control covers :

- People
- Technology
- Policies
- Procedures

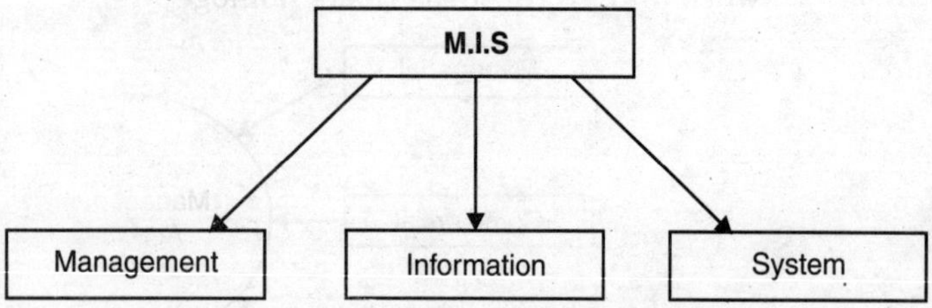

Fig. 2.1. Management Information System

2.2. MANAGEMENT

Management is an art of getting the activities done through the people and with the people. It is normally performed by the managers in an organization.

Operations performed by manager : The operations performed by the manager in an organization are:

- **Plan by making strategies and goals :** Managers plan the work by setting strategies and goals and select the best action to complete the task.

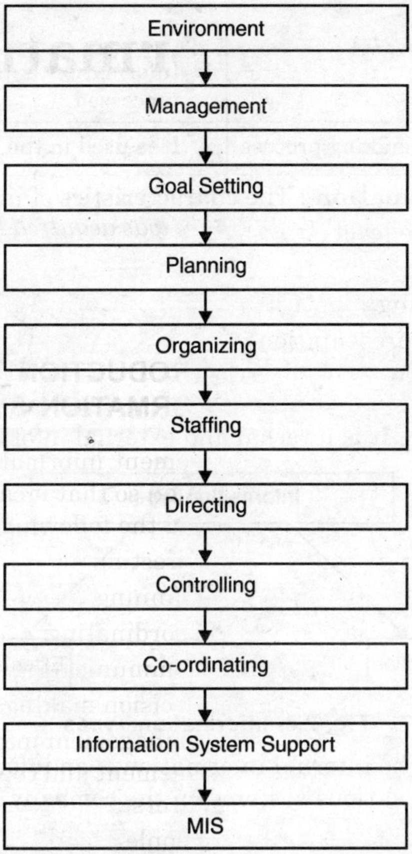

Fig. 2.2. Management Information System

- **Organize the tasks :** Managers organize the tasks that are required to complete the work and group these tasks homogenously.

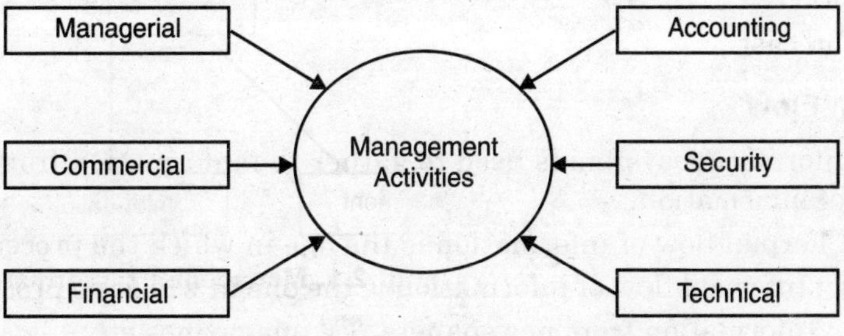

Fig. 2.3. Management Activities

- **Control the performance :** The performance is controlled by the manager by setting some standards and it must be made sure that the standards are not deviated.

DO YOU KNOW?

"To manage is to forecast, to plan, to organize, to command, to co-ordinate and to control."

–Henry Fayol

2.3. INFORMATION

Before knowing what the information is, one should know the difference between data and information.

Data	Information
1. Data is the facts and figures.	**1.** Information is the data that is retrieved and processed.
2. It is not used in the decision making process.	**2.** It is used in the decision making process.

Characteristics of Information : The characteristics of information are :

- Updates the knowledge level
- Reduces uncertainty
- Helps in decision making
- Improves the entity representation

2.3.1. Types of Information

Information is of two types. It is internal and external information.

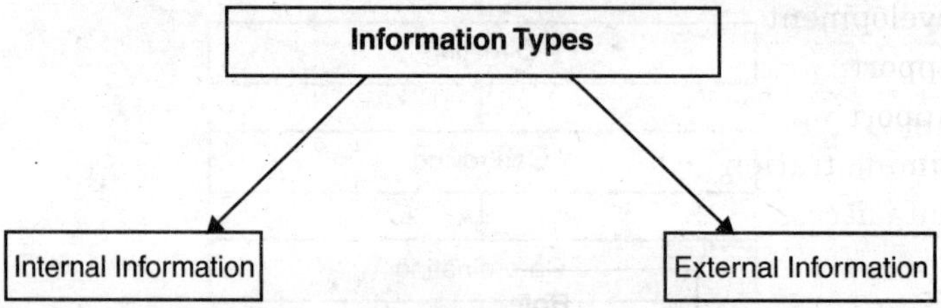

Fig. 2.4. Information Types

Internal Information : The internal information is an information that can be gathered easily. It is gathered from the various systems on the company network. For example: Sales figure.

External Information : The external information is an information that can be gathered by :

- Reading articles
- Economic factors
- Social factors
- Financial forecast

2.3.2. Information Flow

Management information system is used to gather the information from both the formal and informal flow of information.

Formal flow : Formal flow of information is the one in which the procedure is adopted.

Informal flow : Informal flow of information is the one in which no procedure is adopted.

For Example : Information from newspapers, TV, magazines etc.

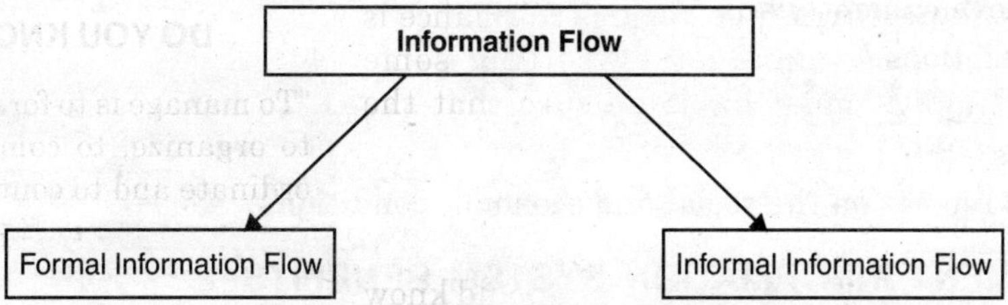

Fig. 2.5. Information Flow

2.4. SYSTEM

A system is a set of elements that work together to achieve a common objective. It helps in optimizing the output of an organization. It helps in connecting the subsystems with the information exchange medium. The objective is :

- Provide synergistic organization in the process.
- Provide information for the decision making process.

These subsystems are part of a larger system. **For example :** Organization is the system and departments like financial department, accounting department etc. are its subsystems.

2.5. ROLE OF MANAGEMENT INFORMATION SYSTEM

The role of a management information system are the following :

- Hardware administration
- Software development
- Software support
- End user support
- Database administration
- Corporate data access

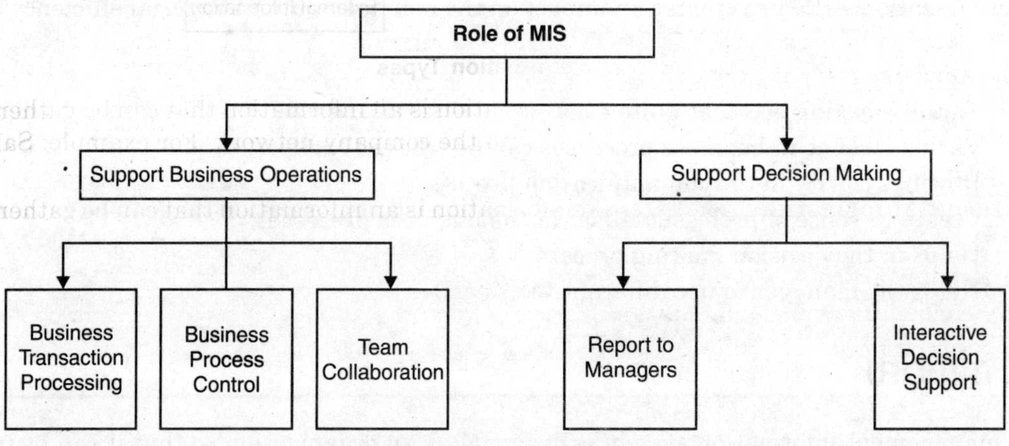

Fig. 2.6. Role of MIS

Also, management information system plays an important role in providing information to the managers. The information includes :

- Number of staff required in each department
- Training requirements

- Career development plans
- Job descriptions
- Budget
- Forecasts
- Progress reports on the social and economic conditions

2.6. MANAGEMENT INFORMATION SYSTEM BENEFITS

The benefits of management information system are :

- It helps in accurate transaction processing.
- It provides managers an interactive support in the decision making process.
- It provides organization an edge over other organizations that do not use a MIS.
- Provides better communication.
- Makes better use of resources.
- Makes better use of time by providing facilities every time.
- Improves customer services and relationships.
- Helps in recording and processing all the business dealings and transactions.
- Provides easy access of information to managers and staff.
- Provides relevant information on the past, present and future events that occurred inside and outside an organization.
- Serves as a system framework for organizing business computer applications.

2.7. IMPACT OF MANAGEMENT INFORMATION SYSTEM

Management information system creates an impact on the :

- Function
- Performance
- Productivity

of an organization. MIS also creates an impact on the management, managerial efficiency and businesses.

The impacts are :

- Target tracking becomes quite easy.
- Various trends in business processes can be tracked easily.
- It brings clarity in the communication process.
- Creates a sense of professionalism in the business processes.
- Helps in the decision making process.
- Helps the manager to use different tools easily.

Summary

➡ A management information system is designed by an organization so that it can be used by the managers in performing functions like directing, planning, coordinating, communicating and decision making.

➡ The management information system is used by the top management and consists of various controls like people, technology, policies and procedures.

- Management is the art of getting the activities done through the people and with the people. It is normally performed by the managers in an organization.
- Planning, organizing and controlling are the three management activities performed by the manager.
- Managers plan the work by setting strategies and goals and select the best action to complete the task.
- Managers organize the tasks that are required to complete the work and group these tasks homogenously.
- The performance is controlled by the manager by setting some standards and it must be noted that the standards should not be deviated.
- Internal and external information are the two types of information.
- The internal information is the information that can be gathered easily. It is gathered from the various systems on the company network.
- The external information is the information that can be gathered by reading articles, economic factors, social factors and financial forecast.
- Management information system is used to gather the information from both formal and informal flow of information.
- Formal flow of information is the one in which procedure is adopted.
- Informal flow of information is the one in which no procedure is adopted.
- A system is a set of elements that work together to achieve a common objective. It helps in optimizing the output of an organization.
- Management information system creates an impact on the function, performance and productivity of an organization.
- Managerial, commercial, financial, accounting, security and technical are the various management activities.

Short Question Answers

1. Define management information system.

Ans. A management information system is designed by an organization so that it can be used by the managers in performing functions like directing, planning, coordinating, communicating and decision making.

2. What are the various controls in a MIS?

Ans. The management information system is used by the top management and consists of various controls like people, technology, policies and procedures.

3. Explain the management concept.

Ans. Management is the art of getting the activities done through the people and with the people. It is normally performed by the managers in an organization.

4. What are the operations performed by a manager?

Ans. Planning, organizing and controlling are the three management activities performed by the manager.

5. State the management activities.

Ans. Managerial, financial, commercial, accounts, technical and security are the management activities.

6. What is the difference between data and information?

Ans.

Data	Information
1. Data is the facts and figures.	1. Information is the data that is retrieved and processed.
2. It is not used in the decision making process.	2. It is used in the decision making process.

7. What are the characteristics of information?

Ans. The characteristics of information are :

- Updates knowledge level
- Reduces uncertainty
- Helps in decision making
- Improves entity representation

8. What are the two types of information?

Ans. Internal and external are the two types of information.

9. Explain internal information.

Ans. The internal information is the information that can be gathered easily. It is gathered from the various systems on the company network. For example: Sales figure.

10. Explain external information.

Ans. The external information is the information that can be gathered by reading articles, economic factors, social factors and financial forecast.

11. What are the two types of information flow?

Ans. Formal and informal are the two types of information flow.

12. Explain formal flow.

Ans. Formal flow of information is the one in which the procedure is adopted.

13. Explain informal flow.

Ans. Informal flow of information is the one in which no procedure is adopted. For Example: Information from newspapers, TV, magazines etc.

14. Explain system concept of MIS.

Ans. A system is a set of elements that work together to achieve a common objective. It helps in optimizing the output of an organization. It helps in connecting the subsystems with the information exchange medium.

15. Explain the role of management information system.

Ans. The role of a management information system is :

- Hardware administration
- Software development
- Software support
- End user support
- Database administration
- Corporate data access

Review Questions

1. Explain MIS in detail.
2. Explain the benefits of MIS in detail.
3. Explain the role of MIS in detail.
4. Explain the impact of MIS in detail.

Chapter 3

Management Information System and The System Approach

"The new information technology... Internet and e-mail... have practically eliminated the physical costs of communications."

Inside this Chapter

- Overview of Decision Support System
- System Approach
- System View of Business
- MIS Organization within the Company
- Management, Information and System Approach

3.1. OVERVIEW OF DECISION SUPPORT SYSTEM

A decision support system is basically an interactive computer based information system that supports the decision making activities of an organization. It serves the :

- Management level
- Operations level
- Planning level

The decision support system includes knowledge based system. It helps the decision makers to compile useful information from :

- Raw data
- Documents
- Business models
- Personal knowledge

3.1.1. History of Decision Support System

- According to Keen, the concept of decision support system came from two areas of research done at the Carnegie Institute of technology.
- In the year 1970, Decision support system became an area of research.
- In 1980s, executive information system, group decision support system and organizational decision support system evolved from model oriented decision support system.

- With the beginning of better technologies, decision support system became an important component of the management design.

3.1.2. Classification of Decision Support System

The decision support system can be classified into five parts. These are :

(*a*) **Communication-driven decision support system :** This type of DSS supports more than one person that is working on a shared task. For example : Groove, Microsoft's NetMeeting.

(*b*) **Data-driven decision support system :** It is also known as data oriented decision support system. It stresses on the access and manipulation of the time series of internal and external data of the company.

(*c*) **Document-driven decision support system :** This type of DSS manages, retrieves and manipulates unstructured information in different electronic formats.

(*d*) **Knowledge-driven decision support system :** This DSS provides specialized expertise in problem solving. It is stored as facts, rules and procedures.

(*e*) **Model-driven decision support system :** This type of DSS makes use of the data and parameters that are provided by the users to help in the decision making process.

3.1.3. Components of Decision Support System

There are three fundamental components of a decision support system. These are :

- **Database :** The database is also known as knowledge base.
- **Model :** The model is for either decision context or the user criteria.
- **User interface :** Users are an important part of the DSS architecture.

3.1.4. Decision Support System Framework

The framework of a decision support system includes :

- People
- Technology
- Development approach

Components : Decision support system consists of :

- **Application :** This must be the actual application that can be used by the user. It helps the decision maker to make decisions about a particular problem.
- **Generator** : The generator consists of an environment that allows people to develop DSS specific applications. Case tools are used by the users. It also consists of special languages, linking modules and function libraries.
- **Tools :** The tools can be lower level hardware or software.

DO YOU KNOW? ✍

The basis of our partnership strategy and our partnership approach. We build the social technology. They provide the music.

–Mark Zuckerberg

The decision support system consists of :

- **Inputs :** Input can be the factors, numbers and characters that can be analysed.
- **Expertise** : Inputs require manual analysis by the user.
- **Outputs** : The output is the transformed data. It helps to generate the decisions of a decision support system.
- **Decisions** : Decisions are the results that are generated by the decision support system. It is created on the basis of user criteria.

3.1.5. Decision Support System Applications

The applications of decision support system are :

- **Clinical decision support system** : It is used for medical diagnosis.
- It is used in business and management.
- It is used in marketing for sustainable development.
- It is used in forest management.

3.1.6. Decision Support System Benefits

The benefits of decision support system are :

- It improves personal efficiency.
- It helps in speeding up the process of decision making.
- It increases organizational control.
- It increases the speed of problem solving in an organization.
- It promotes learning.
- It promotes training.
- It helps in automating the managerial processes.
- It creates a competitive advantage over competition.

3.2. SYSTEM APPROACH

System approach is an approach that is basically designed to make use of scientific analysis in complex organizations. It consists of two operations :

- Developing and managing operating systems.
- Designing information systems for decision making process.

3.2.1. Increased Complexity of Business

The increased business complexity consists of four factors :

(*a*) Technological revolution : The technological revolution consists of :

- Automation techniques are adopted by a large number of industries.
- Industries like transportation, communication, agriculture and manufacturing undergoes a vast change in :
  – Products
  – Techniques
  – Outputs
  – Productivity
- These changes occur at a high speed and the management requires a large number of steps to cope with this change.
- The technological revolution requires managerial revolution.

(*b*) Research and Development : The research and development consists of :

- The technological change is occurring because of the changes in the research and development.
- A large amount of money is being spent upon the research and development.
- Nowadays the products and its operations are becoming quite complex. The lifecycle of the product has become shortened.
- Better planning should be provided.

- Better management should be provided.
- Better information should be provided.

(c) Product changes : The product change consists of :

- Product change has occurred because of the following factors :
  - Advancement in technology
  - Research and development
  - Customer sophistication
- The modern organization has to produce an optimized product in a shorter time.

(d) Information explosion : The information explosion consists of :

- The information explosion has an impact on the :
  - Management
  - Organization
- The modern manager must know the following :
  - Ability to obtain the information
  - Store the information
  - Retrieve the information
  - Displaying of information is very important.

3.2.2. Increased Complexity of Management

The increased management complexity consists of four developments :

(a) Information feedback systems : The information feedback system consists of :

- The information feedback system is concerned with the use of information for the purpose of control.
- It is applied to :
  - Business systems
  - Management systems
  - Engineering systems
  - Biological systems

- The output of the system generates an output. It causes an action that corrects the output thus leading another decision.
- The examples of information feedback systems are :
  - Thermostat-furnace temperature
  - Missile system

(b) Decision making : The decision making consists of :

- The decision making process is a consideration of both the management and information design.
- The decision making process is based upon :
  - Policy
  - Rule
- These are programmed for certain computer applications. Faster and accurate operations are done for these computer applications.
- The examples of decision making are :
  - Payroll
  - Customer billing

– Customer purchasing

(c) Management Science : The management science consists of purpose and tools.

- The purpose of management science is :
  – Design of programmed decision rules.
  – It helps the managers to take complex decisions.
- Simulation tool is used for the construction of a mathematical model for a system that is under study.
- Effects of different management policies, resources etc. can be predicted before the final decision is taken because behavior of the model under manipulation simulates the behavior of the real system.

(d) Electronic computer : The electronic computer consists of :

- The electronic computer helps in the storage of data.
- It helps in processing of information.
- It helps in retrieval of information.
- It helps in performing arithmetic operations.
- A manager considers the electronic computer as a central element in an information system.

The three changes that occur in the management information system are :

- Management becomes more system oriented.
- Management becomes more sophisticated in using management techniques.
- Information is planned.
- Information is available to the managers whenever it is required.

3.3. SYSTEM VIEW OF BUSINESS

Before proceeding with the system view of business, we must know what is a system?

A **system** is basically a set of elements that are joined together to achieve a common objective. These elements form an activity or a scheme.

Business consists of the following activities :

- Improved advertising
- Improved management
- Improved quality control
- Improved design
- Improved credit arrangements
- Improved market strategy

DO YOU KNOW? ✍

We live in a media world that is simultaneously obessed with technology and personality.

–**Mark Zuckerberg**

The system view of business consists of three parameters :

(a) Manufacturing System : The elements of the manufacturing system are :

- People
- Machine
- Facilities

Manufacturing system consists of a set of elements that work to produce :

- Specific number of products
- Types of products

The elements operate on :

- Product specifications

- Schedules
- Raw materials
- Electrical power that is converted into mechanical power.

(*b*) **MIS** : The elements of the MIS are :

- People
- Manual sets
- Data processing equipment

The operations performed by the MIS are :

- Select the information
- Store the information
- Process the information
- Retrieve the information

MIS consists of a set of elements that operate on the data and matter to reduce the uncertainty in the decision making process.

(*c*) **Business organization subsystem :** The element of business organization is people.

The operations performed by the business organization subsystem are :

- Financial
- Design
- Production
- Marketing

The business organization subsystem consists of a set of elements. These elements gather and process the material and information resources. It finds common goals by performing operations on :

- Data
- Energy
- Matter

3.4. MIS ORGANIZATION WITHIN THE COMPANY

The low level organizational information system should be integrated with the upper level organizational information system. It helps to obtain a single management information system for the whole company. It helps in making decision making process at all the levels of the company. The top management is associated with the :

- Design
- Redesign
- Modification
- Maintenance of the entire MIS of the company.

Organization Within the Company Structure

- The management information system is placed under the vice president of the finance department.
- The information systems are designed for controlling the decision services rather than planning the decision services.

The approach followed by the company is :

- The reporting should be done to the president of the company so that information is provided without any problems.

- The reporting can also be done to the corporate head of the computer facilities.

Organization Within the MIS Function

- The system effort should be guided by the :
  - MIS manager
  - Sales manager
  - Production manager
  - Finance manager
  - Engineering manager
- A special team is formed which consists members from :
  - Functional department
  - Technical department
  - Management

  These members work on the design process.
- An organization should be formed to constantly monitor the newly constructed management information system. The role of this organization is to make sure that the MIS works smoothly and is redesigned timely.

3.5. MANAGEMENT, INFORMATION AND SYSTEM APPROACH

The functions of management are classified into following :

- The capabilities of the organization should be matched with the needs of the organization.
- Establishment of a work environment.
- Allocation of resources so that they produce maximum productivity.
- Manage the different responsibilities.

3.5.1. General Management System

The general management system consists of five factors :

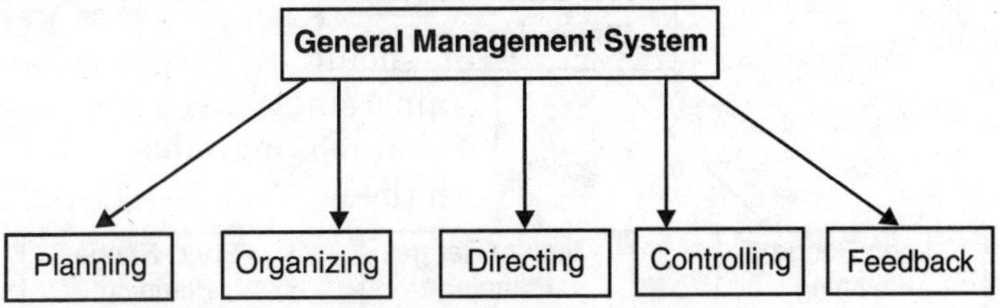

Fig. 3.1. General Management System

(*a*) Planning : The planning consists of :

(*i*) Setting up of objectives

(*ii*) Noticing opportunities

(*iii*) Select the best course of action

(*iv*) Design a program to achieve these objectives

(*b*) Organizing : The organizing consists of :

(*i*) Supervising the action plan

(*ii*) Plan consists of the relationship between the tasks

(*c*) Directing : The directing consists of :

(*i*) Providing directions to the programs
(*ii*) Communicating with the subordinates
(*iii*) Motivating the subordinates

(*d*) **Controlling :** The controlling consists of :
(*i*) Observing the performance
(*ii*) Measuring the performance against the standards
(*iii*) Observing the significant trends
(*iv*) Modification of goals as per requirements

(*e*) **Feedback :** The feedback consists of :
(*i*) Recycling of information
(*ii*) Recycling of actions
(*iii*) Ensure that the objectives are fulfilled

3.5.2. Planning

Planning is the most basic management function. Planning consists of :
- What has to be done?
- When it has to be done?
- How it is to be done?

The planning helps us to reduce the gap between where we are and where we want to go.

Dimensions of Plan : The dimensions of plan can be classified as :
- Long range planning
- Medium range planning
- Short range planning

(*a*) **Long range planning :** The features of long range planning are :
- Operates at the corporate level
- The function is general management
- Companywide scope

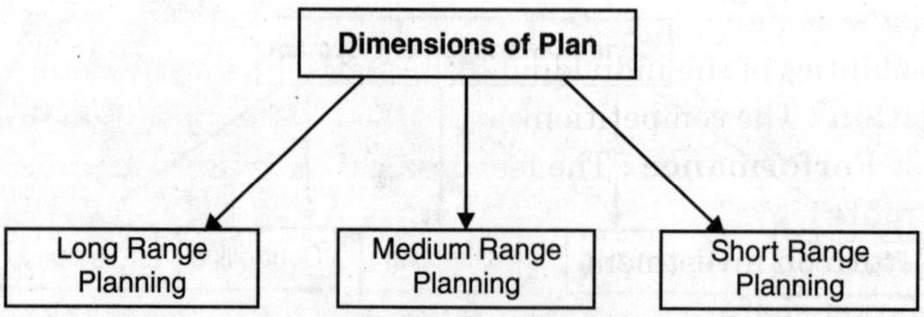

Fig. 3.2. Dimeniosn of Plan

(*b*) **Medium range planning :** The features of medium range planning are :
- Operates at the division level
- The function is marketing
- Business scope

(*c*) **Short range planning :** The features of short range planning are :
- Operates at the operational level
- The function is of scheduling the production process
- Business scope

3.5.3. Information and Planning

The information and planning is classified into three categories :

(a) Environmental Information : The environmental information can be further classified into four parts :

1. **Political and Government** : The features are :
   (*i*) It is required for forecasting the plans.
   (*ii*) It consists of financial policies.
   (*iii*) It consists of tax policies.
   (*iv*) It has an effect on planning decisions.
2. **Economic trends :** The features are :
   (*i*) Employment
   (*ii*) Productivity
   (*iii*) Capital investment
   (*iv*) Price
   (*v*) Wage levels
3. **Technological Environment :** The features are :
   (*i*) Forecast the technological changes.
4. **Factors of production** : The features are :
   (*i*) Cost
   (*ii*) Labor
   (*iii*) Location
   (*iv*) Availability

(b) Competitive Information : The competitive information can be further classified into three parts :

1. **Industry demand :** The features are
   (*i*) Forecasting of demands by the firm
2. **Firm demand :** The firm demand is the function of the :
   (*i*) Industry demand
   (*ii*) Capabilities of the individual firm relative to the competing firm.
3. **Competition :** The competition can further be classified into three parts :
   (a) Past Performance : The features are :
   - Profit
   - Return on investment
   - Market share
   - Helps to identify competition

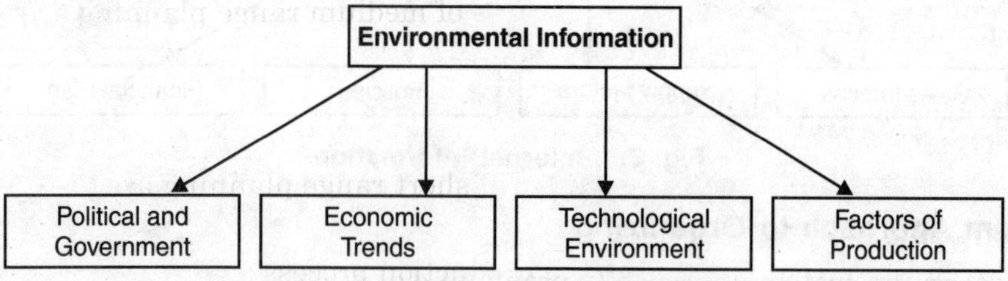

Fig. 3.3. Environmental Information

(*b*) **Present Performance :** The features are :
- Price
- Strategy
- Advertisement

(*c*) **Future Performance :** The features are :
- Information regarding new products
- Research and development

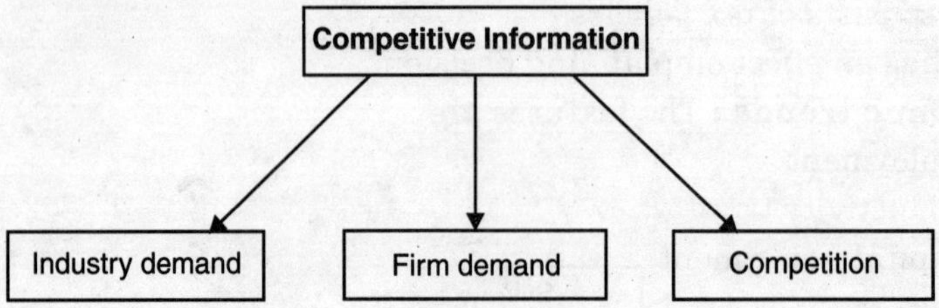

Fig. 3.4. Competitive Information

(*c*) **Internal Information :** The internal information can be further classified into four parts :

(*i*) **Sales forecast :** The features are :
- Important planning document

(*ii*) **Financial Plan :** The features are :
- Involves the entire organization
- Provides information for sub-plans

(*iii*) **Supply factors :** The features are :
- Labor
- Capital
- Organization
- Equipment

(*iv*) **Policies :** The features are :
- Product
- Marketing
- Financial policies
- Term policies

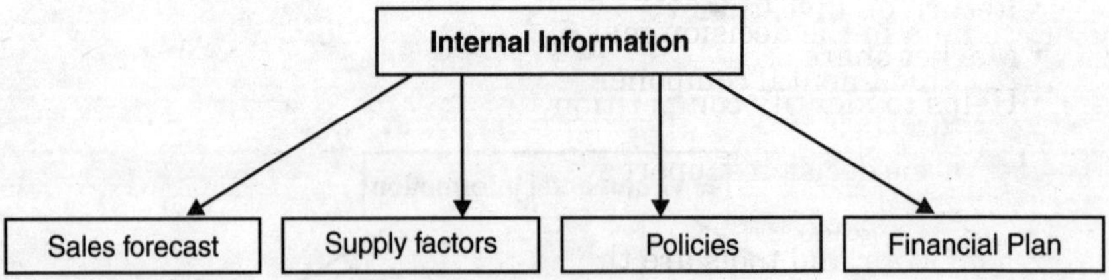

Fig. 3.5. Internal Information

3.5.4. System Approach to Organizing

The phases in the system approach to organizing are :
- Achievement

- Composition
- Linking
- Result

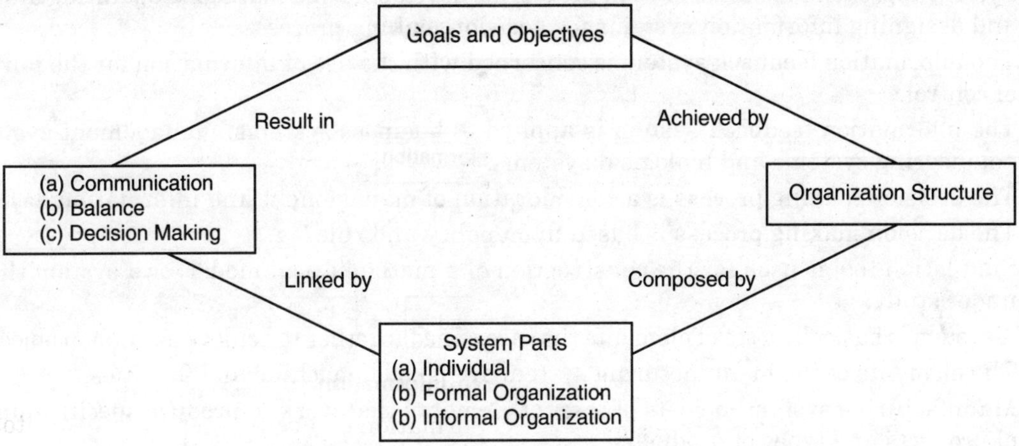

Fig. 3.6. System approach to organizing

Summary

- A decision support system is basically an interactive computer based information system that supports the decision making activities of an organization.
- Decision support system serves the management level, operation level and planning level.
- Communication-driven decision support system supports more than one person that is working on a shared task.
- Data-driven decision support system stresses on the access and manipulation of the time series of internal and external data of the company.
- Document-driven decision support system manages, retrieves and manipulates unstructured information in different electronic formats.
- Knowledge-driven decision support system provides specialized expertise in problem solving. It is stored as facts, rules and procedures.
- Model-driven decision support system makes use of data and parameters that are provided by the users to help in the decision making process.
- There are three fundamental components of decision support system. These are database, model and user interface.
- The framework of the decision support system includes people, technology and development approach.
- Application, generator and tools are the components of decision support system.
- The generator consists of an environment that allows people to develop DSS specific applications.
- The tools can be lower level hardware or software.
- The decision support system consists of inputs, outputs, expertise and decisions.
- Decisions are the results that are generated by the decision support system. It is created on the basis of user criteria.

- Input can be factors, numbers and characters that can be analysed.
- System approach is an approach that is basically designed to make use of the scientific analysis in complex organizations.
- System approach consists of two operations i.e. developing and managing operating systems and designing information systems for decision making process.
- The information feedback system is concerned with the use of information for the purpose of control.
- The information feedback system is applied to business systems, management systems, engineering systems and biological systems.
- The decision making process is a consideration of management and information design.
- The decision making process is based upon policy and rule.
- Simulation tool is used for the construction of a mathematical model for a system that is under study.
- A system is basically a set of elements that are joined together to achieve a common objective.
- The elements of the manufacturing system are people, machine and facilities.
- Manufacturing system consists of a set of elements that work to produce specific number of products and types of products.
- The elements of the MIS are people, manual sets and data processing equipment.
- The operations performed by the MIS are selecting the information, storing the information, processing the information and retrieving the information.
- MIS consists of a set of elements that operates on the data and matter to reduce the uncertainty in the decision making process.
- The operations performed by the business organization subsystem are financial, design, production and marketing.
- The business organization subsystem consists of a set of elements. These elements gather and process the material and information resources.
- The top management is associated with the design, redesign, modification and maintenance of the entire MIS of the company.
- The management information system is placed under the vice president of the finance department.
- The information systems are designed for controlling the decision services rather than planning the decision services.
- The planning consists of setting up of objectives, noticing opportunities, select the best course of action and design a program to achieve these objectives.
- The organizing consists of supervising the action plan. The plan consists of the relationship between the tasks.
- The directing consists of providing directions to the programs, communicating with the subordinates and motivating the subordinates.
- The controlling consists of observing the performance, measuring the performance against the standards, observing the significant trends and modification of goals as per requirements.
- The feedback consists of recycling of information, recycling of actions and ensures that the objectives are fulfilled.
- The dimensions of plan can be classified as long range planning, medium range planning and short range planning.

Short Question Answers

1. What is decision support system?

Ans. A decision support system is basically an interactive computer based information system that supports the decision making activities of an organization. It serves the management level, operations level and planning level.

2. State the history of decision support system.

Ans. The history of decision support system is :

- According to Keen, the concept of decision support system came from two areas of research done at the Carnegie Institute of technology.
- In the year 1970, Decision support system became an area of research.
- In 1980s, executive information system, group decision support system and organizational decision support system evolved from model oriented decision support system.
- With the beginning of better technologies, decision support system became an important component of the management design.

3. Classify the decision support system.

Ans. The decision support system can be classified into five parts. These are :

(*a*) **Communication-driven decision support system :** This type of DSS supports more than one person that is working on a shared task. For example : Groove, Microsoft's NetMeeting.

(*b*) **Data-driven decision support system :** It is also known as data oriented decision support system. It stresses on the access and manipulation of the time series of internal and external data of the company.

(*c*) **Document-driven decision support system :** This type of DSS manages, retrieves and manipulates unstructured information in different electronic formats.

(*d*) **Knowledge-driven decision support system :** This DSS provides specialized expertise in problem solving. It is stored as facts, rules and procedures.

(*e*) **Model-driven decision support system :** This type of DSS makes use of data and parameters that are provided by the users to help in the decision making process.

4. What are the components of DSS?

Ans. There are three fundamental components of decision support system. These are :

- **Database :** The database is also known as knowledge base.
- **Model :** The model is for either decision context or the user criteria.
- **User interface :** Users are an important part of the DSS architecture.

5. Explain DSS framework.

Ans. The framework of decision support system includes :

- People
- Technology
- Development approach

6. Explain the applications of DSS.

Ans. The applications of decision support system are :

- It is used for medical diagnosis.
- Used in business and management.
- It is used in marketing for sustainable development.
- It is used in forest management.

7. What are the benefits of DSS?

Ans. The benefits of decision support system are :

- It improves personal efficiency.

- It helps in speeding up the process of decision making.
- It increases organizational control.
- It increases the speed of problem solving in an organization.
- It promotes learning.
- It promotes training.
- It helps in automating the managerial processes.
- It creates a competitive advantage over competition.

8. Explain system approach.

Ans. System approach is an approach that is basically designed to make use of scientific analysis in complex organizations. It consists of two operations :

- Developing and managing operating systems.
- Designing information systems for decision making process.

9. Explain increased complexity of business.

Ans. The increased business complexity consists of four factors :

- Technological revolution
- Research and development
- Product changes
- Information explosion

10. Explain technological revolution.

Ans. The technological revolution consists of :

- Automation techniques are adopted by a large number of industries.
- Industries like transportation, communication, agriculture and manufacturing undergo a vast change in products, techniques, outputs and productivity.
- These changes occur at a high speed and the management requires a large number of steps to cope up with this change.
- The technological revolution requires managerial revolution.

11. Explain research and development.

Ans. The research and development consists of :

- The technological change is occurring because of the changes in the research and development.
- A large amount of money is being spent upon the research and development.
- Nowadays the products and its operations are becoming quite complex. The lifecycle of the product has become shortened.
- Better planning should be provided.
- Better management should be provided.
- Better information should be provided.

12. Explain product change.

Ans. The product change consists of :

- Product change has occurred because of the following factors i.e. advancement in technology, research and development and customer sophistication.
- The modern organization has to produce an optimized product in a shorter time.

13. Explain information explosion.

Ans. The information explosion consists of :

- The information explosion has an impact on the management and organization.
- The modern manager must know that ability to obtain the information, store the information, retrieve the information and display the information is very important.

14. Explain increased management complexity.

Ans. The increased management complexity consists of four developments :

- Information feedback system
- Management science
- Electronic computer
- Decision making

15. Explain information feedback systems.

Ans. The information feedback system consists of :

- The information feedback system is concerned with the use of information for the purpose of control.
- It is applied to business systems, management systems, engineering systems and biological systems.
- The output of the system generates an output. It causes an action that corrects the output thus leading another decision.

16. Explain decision making.

Ans. The decision making consists of :

- The decision making process is a consideration of management and information design.
- The decision making process is based upon policy and rule.
- These are programmed for certain computer applications. Faster and accurate operations are done for these computer applications.

17. Explain management science.

Ans. The management science consists of purpose and tools.

- The purpose of management science is the design of programmed decision rules.
- It helps the managers to take complex decisions.
- Simulation tool is used for the construction of a mathematical model for a system that is under study.
- Effects of different management policies, resources etc. can be predicted before the final decision because behaviour of the model under manipulation simulates the behaviour of the real system.

18. Explain electronic computer.

Ans. The electronic computer consists of :

- The electronic computer helps in storage of data.
- It helps in processing of information.
- It helps in retrieval of information.
- It helps in performing arithmetic operations.
- A manager considers the electronic computer as a central element in an information system.

19. What are the changes that occur in MIS?

Ans. The three changes that occur in the management information system are :

- Management becomes more system oriented.
- Management becomes more sophisticated in using management techniques.
- Information is planned.
- Information is available to the managers whenever required.

20. What is a system?

Ans. A system is basically a set of elements that are joined together to achieve a common objective. These elements form an activity or a scheme.

21. What are the activities in a business?

Ans. Business consists of the following activities :

- Improved advertising
- Improved management

- Improved quality control
- Improved design
- Improved credit arrangements
- Improved market strategy

22. What are the elements of a manufacturing system?

Ans. The elements of the manufacturing system are people, machine and facilities.

23. What are the operations performed by the elements of a manufacturing system?

Ans. The elements operate on product specifications, schedules, raw materials and electrical power that is converted into mechanical power.

24. What are the elements of a MIS?

Ans. The elements of the MIS are people, manual sets and data processing equipment.

25. What are the operations performed by the management information system?

Ans. The operations performed by the MIS are selecting the information, storing the information, processing the information and retrieving the information.

26. What are the elements of a business organization subsystem?

Ans. The element of business organization is people.

27. What are the operations performed by a business organization subsystem?

Ans. The operations performed by the business organization subsystem are financial, design, production and marketing.

28. Explain the organization within the company structure.

Ans. The organization within the company structure can be explained as :

- The management information system is placed under the vice president of the finance department.
- The information systems are designed for controlling the decision services rather than planning the decision services.

29. Explain organization within the MIS function.

Ans. The organization within the MIS function can be explained as :

- The system effort should be guided by the MIS manager, sales manager, production manager, finance manager and engineering manager.
- A special team is formed which consists of members from functional department, technical department and management.
- An organization should be formed to constantly monitor the newly constructed management information system. The role of this organization is to make sure that the MIS works smoothly and is redesigned timely.

30. Explain planning.

Ans. The planning consists of setting up of objectives, noticing opportunities, select the best course of action and design a program to achieve these objectives.

31. Explain Organizing.

Ans. The organizing consists of supervising the action plan. Plan consists of the relationship between the tasks.

32. Explain directing.

Ans. The directing consists of providing directions to the programs, communicating with the subordinates and motivating the subordinates.

33. Explain controlling.

Ans. The controlling consists of observing the performance, measuring the performance against the standards, observing significant trends and modification of goals as per requirements.

34. Explain feedback.

Ans. The feedback consists of recycling of information and actions. It must ensure that the objectives are fulfilled.

35. Explain dimensions of plan.

Ans. The dimensions of plan can be classified as long range planning, medium range planning and short range planning.

36. Explain the features of political and government.

Ans. The features are :

- It is required for forecasting the plans.
- It consists of financial policies.
- It consists of tax policies.
- It has an effect on planning decisions.

37. Explain the features of economic trends.

Ans. The features are :

- Employment
- Productivity
- Capital investment
- Price
- Wage levels

38. Explain the features of technological environment.

Ans. The features are :

- Forecast the technological changes

39. Explain the factors of production.

Ans. The features are :

- Cost
- Labor
- Location
- Availability

40. Classify competitive information.

Ans. Industry demand, firm demand and competition are the classifications of competitive information.

Review Questions

1. Explain decision support system in detail.
2. Explain system approach in detail.
3. Explain system view of business in detail.
4. Explain MIS organization within the company.
5. Explain management, information and system approach in detail.

□□□

Chapter 4

Manager Overview

"We may ask for information, but we are usually only interested in what confirms our opinions."

Inside this Chapter

- Managerial Overview
- Information Manager
- Business Process Re-engineering
- Business Information Manager

4.1. MANAGERIAL OVERVIEW

Before understanding the manager overview, it is important to first understand who is an information system manager. So, an information system manager is a person who is responsible for :

- Manage the computer systems present in a company
- Installation processes
- To make sure that the back-up systems works in an efficient way
- To purchase the required hardware and software
- Participate in the organizational policy making
- Takes care of the quality standard

Information managers are seen in every organization. Technicians, programmers and database-manager report to the information manager.

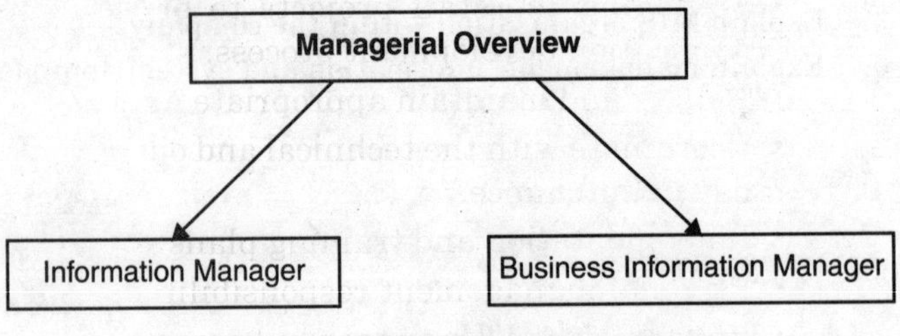

Fig. 4.1. Managerial Overview

4.2. INFORMATION MANAGER

The major responsibilities of an information manager are :

- Implementation of technology in an organization
- Overview the work of the various computer personnel
- Evaluating the user needs
- Evaluate the functionality of the system
- Planning, developing and implementing the information system
- Scheduling upgrade
- Installation of new systems
- Ensure that the software laws are not broken
- Provision of secure network access to the remote users
- Maintaining the security of data from both external and internal attack
- Providing appropriate support to the users
- Remaining up to date with the new technologies

4.3. BUSINESS PROCESS RE-ENGINEERING

The companies that undergo business process engineering deals with the information system manager for change management. **Now the question is what is business process re-engineering**.

Business process re-engineering deals with the re-designing of the way the work is done so that the organization can meet its goals easily in a less price. So it is the responsibility of the information manager to :

- Understanding the capabilities of the system
- Understanding the limitations of the system in terms of cost and technology
- Ensure that the process of training goes in a proper way.

DO YOU KNOW?

The opportunity to engage people who are very far away from each other and yet might have interesting information is better here than has ever been possible.

4.4. BUSINESS INFORMATION MANAGER

The business process manager is a person that is a member of the Information Technology management team. The responsibilities are :

- Managing the requirements of the client
- Maintenance of Information Technology's
- Maintenance of project activity
- Setting up of appropriate service level agreements
- Managing the staff projects to bring client business requirements through the application development process.
- Deliver and maintain appropriate application systems
- Coordinate with the technical and operation group for the activity that requires day to day maintenance.
- Implementation and training plans
- Personal management responsibility for business process information systems
- Providing the IT management committee with an estimate of the scope of the service that is to be provided.

Summary

- An information system manager is a person who is responsible for managing the computer systems in a company.
- Information managers are seen in every organization.
- Technicians, programmers and database-manager report to the information manager.
- Manager overview can be classified as information managers and business information managers.
- Information manager is responsible for planning, developing and implementing an information system.
- The companies that undergo business process engineering deals with the information system manager for change management.
- Business process re-engineering deals with the re-designing of the way the work is done so that the organization can meet its goals easily in a less price.
- The business process manager is a person that is a member of the Information Technology management team.

Short Question Answers

1. Who is an information manager?

Ans. An information system manager is a person who is responsible for managing the computer systems in a company. Information managers are seen in every organization. Technicians, programmers and database manager report to the information manager.

2. Classify manager's overview.

Ans. Manager's overview can be classified as information managers and business information managers.

3. State few responsibilities of an information manager.

Ans. The major responsibilities of an information manager are:

- Implementation of technology in an organization
- Overview the work of the various computer personnel
- Evaluating the user needs
- Evaluate the functionality of the system
- Planning, developing and implementing the information system
- Remaining up to date with the new technologies

4. What is business process re-engineering?

Ans. The companies that undergo business process engineering deals with the information system manager for change management. Business process re-engineering deals with the re-designing of the way the work is done so that an organization can meet its goals easily in a less price.

5. What is the responsibility of an information manager in business process re-engineering?

Ans. The responsibility of the information manager is :

- Understanding the capabilities of the system
- Understanding the limitations of the system in terms of cost and technology
- Ensure that the process of training goes in a proper way

6. Who is a business information manager?

Ans. The business process manager is a person that is a member of the Information Technology management team.

7. What are the responsibilities of a business information manager?

Ans. The responsibilities of a business information manager are :

- Managing the requirements of the client
- Maintenance of Information Technology's
- Maintenance of project activity
- Setting up of appropriate service level agreements
- Managing the staff projects to bring client business requirements through the application development process.
- Deliver and maintain appropriate application systems
- Coordinate with the technical and operation group for the activity of day to day maintenance.
- Implementation and training plans

Review Questions

1. Explain the manager's overview in detail.
2. Explain the responsibilities of the information manager.
3. Explain the responsibilities of the business information manager.
4. Explain business process re-engineering in detail.

□□□

Chapter 5

Information Technology

"Some say Google is God. Others say Google is Satan. But if they think Google is too powerful, remember that with search engines unlike other companies, all it takes is a single click to go to another search engine."

Inside this Chapter

5.1. COMPUTER HARDWARE

The **hardware** is the physical parts of the computer. The computer hardware can be classified as :

(*a*) **Motherboard :** The motherboard of the computer is the central circuit board that makes up the computer system. It holds up :
- CPU
- Main memory
- Slots for expansion cards

(*b*) **Storage :** Storage of a computer system means those parts of the computer system that stores data for some period of time.

(*c*) **Input-Output devices :** The input-output device refers to the various interfaces that different functional units of system use to communicate with each other.

(*d*) **Power supply :** It is a case that holds :
- Transformer
- Voltage control
- Fan

(*e*) **Interface controller :** Interface controller like USB, firewire is used to connect the computer to external peripheral devices like printers, scanners etc.

5.2. COMPUTER SOFTWARE

The **computer software** is a collection of computer programs. In contrast to the computer hardware, the software cannot be touched. These programs provide instructions to the computer for what and how to do. Thus, software is a set of :

- Programs
- Procedures
- Algorithms

The functions are performed either directly or indirectly. It is done directly by providing instructions to the computer hardware. It is performed indirectly by serving as an input to other software.

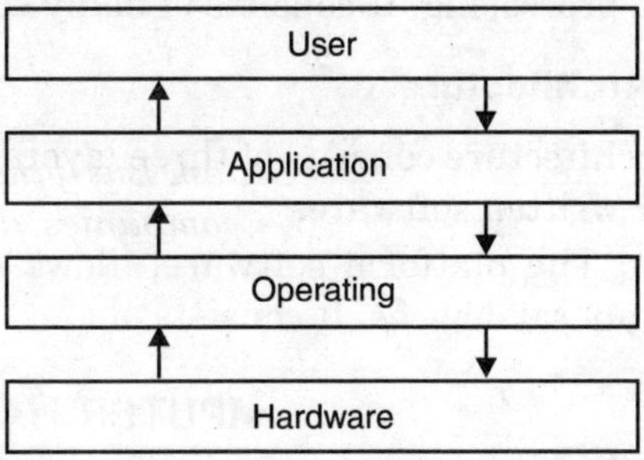

Fig. 5.1. Computer Hardware and Software

System Software : System software is responsible for controlling, integrating and managing the individual hardware components of a computer system. The users are not concerned with the low level details like transferring of data from the memory to disk.

Components of System Software : The components of system software are :

- Device drivers
- Operating systems
- Servers
- Window systems
- Management tools
- Networking control software
- Disk formatters
- File managers

> **DO YOU KNOW?**
>
> It has become applied obvious that our technology has exceeded our humanity.
>
> **–Albert Einstein**

Functions of system software : The functions performed by the system software are :

- It is responsible for managing independent hardware components.
- Help the application software programmer from the complex details of the system.
- Partition the computer resources in a safe and stable manner.

For Example : Microsoft word is developed in languages like C, C++.

5.2.1. Types of Software

The classification of software is done in two parts :

(*a*) **Programming Software :** The programming software consists of tools in the form of programs and applications. The software developer makes use of the programming software to perform the following functions :

- Create the programs

- Debug the programs
- Maintain the programs
- Support other programs

For example : The examples of programming software are :

- Compilers
- Debuggers
- Interpreters
- Linkers
- Text editors

(*b*) **Application Software :** The application software is basically a set of programs that performs specific data processing. It consists of many softwares.

5.2.2. Computer Software Architecture

The computer software architecture consists of three layers. These are platform software, application software and user written software.

(*a*) **Platform Software :** The platform software allows the user to interact with the computer and its peripheral devices. It consists of :

- Firmware
- Device drivers
- Operating system
- Graphical user interface

(*b*) **Application Software :** Application software are the independent programs that are modified according to the platforms. They are purchased separately from the computer hardware. For Example :

- Microsoft office suite
- Video games

(*c*) **User-written software :** User written software is the software that is created by the user. For Example :

- Spreadsheet templates
- Word processor templates

5.3. CONVERSION OF MANUAL TO COMPUTER BASED SYSTEM

The conversion of computer to manual based computer system is used to understand the computer based management information system.

The steps involved in the conversion to computer based system are :

(*a*) **System Description :** The system description is prepared after the definition of the problems. System description consist of statements related to the :

- Input of the system
- Output of the system
- Operations performed
- Files required

The purpose of system description is :

- Logical information flow
- Logical operations

These operations are used to carry out the particular system design.

Forms of System Description : There are two forms of system description. These are :

1. **Narrative system description :** In the narrative system description, the operations of the system are described in English language. The description must help the user and technicians to understand the operations of the system. The narrative system description must consist description about the :
   (*a*) Input
   (*b*) Output
   (*c*) Files
   (*d*) Operations
2. **Pictorial system description :** Pictorial system description provides a quick analysis on the work being performed by the system. It provides a visual of the entire operation. **For example :** System flowchart.

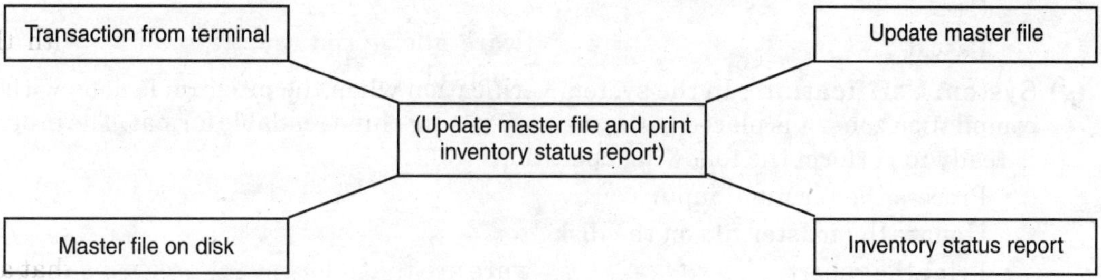

Fig. 5.2. Conversion Process

(*b*) **Input Documents :** The input documents are the documents that are required to specify the method to put the information in a form that is acceptable to the computer. The computer acceptable form is the machine usable form. The input format is selected on the basis of :

- Volume of information
- Frequency
- Accuracy
- Verification

In the input documents, exact layout is necessary because computer program consists of precise sequence of steps. It operates only if the data is present in its desired position.

(*c*) **Output Documents :** The output documents show the purpose of the entire operations that were performed on the input data. The management is concerned with the output data.

(*d*) **File Design :** The file design consists of the logic that is required to control the flow of data through the system. It is a part of the system design only. It should be noted that the flow of data depends solely on the design of the data files. In the file design, following factors should be considered :

- Equipment
- Storage
- Capacity
- Input media

- Output media
- Format

(e) Program Logic : In the program logic, program flowcharts are used. The program logic is a step by step representation of the method to complete the task. It deals with the entire task in a systematic manner. It is used to perform the following functions :

- Organize facts
- Outline problems
- Outline logics
- Outline solutions

(f) Computer Programming : The computer programming is concerned with the method to instruct the processor on performing the operations on the input data so that the output is produced in the required format.

It consists of different programming languages like :

- C
- C++
- Pascal

(g) System Verification : In the system verification when the program is done with the compilation zone, it is placed in the memory in a machine readable format. The program is ready to perform the following operations :

- Process the terminal input
- Update the master file on the disk
- Print the report

(g) Documentation : The documentation process is the most important concept. The documentation process is of three types.

1. **Input :** The documentation should consist of the simple description about the system. It should provide a clear description of the required inputs in the system.
2. **Maintenance :** The documentation process should cover the technical documentation.
3. **Output :** The documentation should consist of the simple description about the system. It should provide a clear description of the outputs and their limitations.

5.4. COMPONENTS OF A COMPUTER SYSTEM

The components of a computer system are :

A. Input : Input is the data that is required to enter into the computer system. The input data should be present in machine readable format. The data that is received from punch cards, magnetic tapes, keyboards are converted into machine readable format.

B. Central Processor : The central processor is the main part of the computer. It consists of three parts. These are :

- **Control Section :** The control section is responsible for integrating all the operations of the computer system. It basically directs and coordinates the operations.
- **Control input-output devices :** The control of input and output devices is associated with two factors. First is the entry and removal of the information from the storage devices. Second is the routing of information between the storage devices and arithmetic logic section.

- **Arithmetic logic section** : The arithmetic logic section performs the arithmetic and logic operations. The arithmetic operation is the calculation process and shifting of numbers. The logic operation is the decision to change the sequence of execution of the instructions.

C. **Storage** : The data should be placed in the storage devices before performing the processing functions on the data. Storage is of two type's i.e. internal and external storage.

Internal storage : Internal storage is known as memory. It allows the machine to store data in electronic form. It allows the machine to store long instructions that is required to guide the machines regarding what operations are to be performed.

External storage : External storage is further classified into two types. Direct access and sequential are the two types of external storage.

- **Direct access** : In the direct access, the data can be read from anywhere. It is not necessary that the required data should be read from the beginning of the file. For Example : disk and magnetic drums.
- **Sequential** : The sequential storage requires that the data must be read from the beginning of the file so that the desired record is read and written. For Example : magnetic tapes.

D. **Output** : The output is the final result of the data processing. The output is recorded on different media's like cards and magnetic media. The record information can either be updated inventory master file or inventory status report.

5.5. DATA BASE MANAGEMENT SYSTEM

A database is a collection of data records, files and objects. A data base management system is a software package that :

- Creates the database
- Maintains the database
- Use the database

It can be accessed by multiple users and multiple application programmers at the same time.

5.5.1. Components of a Database

The components of a database are :

- **DBMS Engine** : It accepts requests from other database systems and converts them into physical parts. It accesses the database and data dictionary.
- **Data definition subsystem** : It helps the user to create the data dictionary and define the structure of the files.
- **Data manipulation subsystem** : As the name suggests, it helps the user to manipulate the system. It helps the user to add, modify and delete the information from the database.
- **Application generation subsystem** : It helps the user to develop applications. The user should perform a detailed series of tasks to process a transaction.
- **Data administration subsystem** : It helps the user to manage the entire database by providing facilities like backup and recovery, security management etc.

5.5.2. Features of Database

The features of the database are listed below :

(*a*) **Support large amount of data :** The DBMS is designed so that it can support large amount of data. There are methods to store and manipulate the data. For Example : oracle can store terabytes of data.

(*b*) **Data Sharing :** DBMS allows data to be shared between two or more users.

(*c*) **Data Concurrency :** DBMS allows data to be accessed by multiple users at the same time.

(*d*) **Data Locking :** DBMS provides the facility of locking the data to avoid the problem of two users locking the data at the same time. This method is an automatic and transparent process.

(*e*) **Data Security :** DBMS makes sure that the data is accessed by authorized users. It is by default that the data of a user cannot be accessed by another user until the user gives permission of accessing the data.

(*f*) **Data Integrity :** The integrity of the data should be maintained. DBMS provides methods to maintain the integrity of the data. There are three integrity rules : domain, entity and referential.

(*g*) **Fault Tolerance :** DBMS allows fault tolerance. It allows the user to correct the mistakes in a mean time.

(*h*) **Fault Recovery :** DBMS allows recovery in the case of failure.

(*i*) **Language Support :** DBMS supports data access and manipulation languages. For example : SQL.

5.5.3. Database Models

There are five database models which are listed in a detailed manner below :

(*a*) **Hierarchical Model :** The hierarchical database model is a model in which the data is organized in a tree like form. The hierarchical model is a simple model. It represents information using one to many relationships. **For example :** Storing geographical information, mobile etc.

(*b*) **Network Structure :** The network structure is a database model of representing objects and their relationships. It is a flexible model. The network structure implements many to many relationships.

(*c*) **Relational Structure :** Relational structure is a database model that makes use of two dimensional rows and columns to store the data. It is a complex method and is not suitable for the end user. The relational structure is used by the mainframe, midrange and microcomputer systems.

(*d*) **Multidimensional Structure :** The multidimensional structure is very similar to the relational model. The multidimensional structure is quite easy to maintain and understand. The records are stored as fundamental attributes.

(*e*) **Object-oriented structure :** The object oriented structure is popular for the multimedia. It has an ability to handle graphics, pictures, voice and text. It is designed to work with the object oriented programming languages like java.

5.5.4. Management and Database

There is a relationship between the management of an organization and its database.

- Managers are dependent on the database for retrieving and processing the data so that useful information can be produced.

- Good managers pay special attention to the database.
- Manager does not administer the database at operational level.
- Manager defines, organizes and uses the database in a correct manner.

5.5.5. Advantages of Database

The advantages of database are :

- It keeps a check on the data redundancy.
- It ensures data integrity.
- It ensures that the data can be shared across all applications.
- It contains a backup and recovery procedure of data.
- It provides better services to the users.
- Improved flexibility.
- Cost of developing and maintaining the system is low.
- Improved security.

5.5.6. Disadvantages of Database

The disadvantages of using database are :

- Implementation of a database can be expensive.
- Implementation of a database can be time consuming.
- Even with the provision of high security, it is possible for unauthorized users to access the database.

5.6. RELATIONAL DATABASE MANAGEMENT SYSTEM

A relational database management system is a database management system that is based on a relational model. It is designed by **E.F. Codd**. It is based on the theory of sets and relations. It is the most popular database.

- Relational model represents data in a table. The table contains an array of rows and columns.
- The row contains data related to an entity.
- The column contains data related to a single attribute of the entity.

Advantages

- It is simple in nature.
- It is easy to understand.
- It is easy to manipulate.
- It allows to have a dynamic relationship between the entities.

5.7. DATA MINING

Data mining is a tool that is used for analysing large amount of data that is present in a data warehouse. It helps in decision making process. It is also known as "**Knowledge discovery of database**". The goal of the data mining process is to extract knowledge from the existing data and transform into a human understandable format so that it can be used further.

5.7.1. Data Mining Techniques

The most common used techniques in data mining are :

(*a*) **Case-based reasoning :** This uses historical cases to recognize patterns.

(*b*) **Neural computing :** Neural computing is a machine based learning approach. The historical data is examined for pattern recognition. These patterns are used for decision support.

(*c*) **Intelligent agents :** It is an approach for retrieving information from the Internet. A large amount of information is available on the Internet.

(*d*) **Other Tools :** Many tools like decision trees, genetic algorithms are used in the data mining process.

5.7.2. Information Types

The information types can be classified into five types. These are :

(*a*) **Classification :** Define characteristics of certain groups.

(*b*) **Clustering :** Clustering identifies a group of items that share a particular characteristic.

(*c*) **Association :** Association identifies relationship between events that occur at one time.

(*d*) **Sequencing :** Sequencing is quite similar to association. The relationship exists over a period of time.

(*e*) **Forecasting :** Forecasting eliminates the values that are based upon patterns within large sets of data.

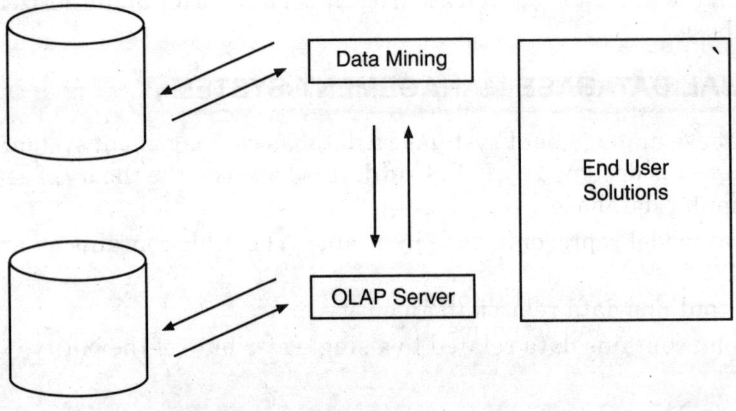

Fig. 5.3. Data Mining Architecture

5.7.3. Applications of Data Mining

There is a large variety of the applications of data mining. These are :

(*a*) **Retailing and sales :** It is used in predicting sales and determining the correct inventory levels.

(*b*) **Banking :** It helps in forecasting :

- Fraud credit card transactions
- Kind of customers responding to loans

(*c*) **Manufacturing and Production :** It helps in predicting :

- Machine failures
- Factors controlling the optimization of manufacturing capacity

(*d*) **Insurance :** It helps in predicting :
- Medical coverage costs
- Claim the amounts

(*e*) **Computer hardware and software :** It helps in predicting :
- Disk drive failures
- Time taken to create new chips

5.7.4. Capabilities of Data Mining

There are some advantages of using data mining that are listed below :

(*a*) **Automated prediction of trends and behaviours :** Data mining automates the process of finding predictive information in large databases. For example : targeted marketing, forecasting bankruptcy.

(*b*) **Automated discovery of previous unknown patterns :** Data mining identifies previous unknown data in a single step.

(*c*) They analyse massive database in a fraction of minutes.

(*d*) Data mining procedure can be easily conducted by non-programmers.

5.7.5. Data Mining Failure

The reasons for data mining failure are :
- Inappropriate architecture
- Lack of training
- Poor use of technology
- Lack of data quality
- Lack of user input
- Interference of corporate politics

5.8. DATA WAREHOUSING

Data warehouse is a repository of information that is organized so that it can be accessed easily for analytical processing activities. Data warehouse is a collection of information that is obtained from either :
- Operational systems
- External data sources

5.8.1. Purpose of Data Warehousing

The purpose of data warehousing is :
- **Realize the value of data :** The realization of data is done by the methods of reporting and analysing.
- Better decision making
- Convert data into information
- Create a sense of competition

5.8.2. Components of Data Warehousing

There are three components of data warehousing. These are :

(*a*) **Staging Area :** Staging area is a repository where the transaction data can be transferred for and in the data warehouse.

(*b*) **Data Mart :** The data warehouse is an union of a set of data marts.

(*c*) **Operational Data Store :** The operational data store is modelled to support the real time reporting needs.

5.8.3. Data Warehousing Framework

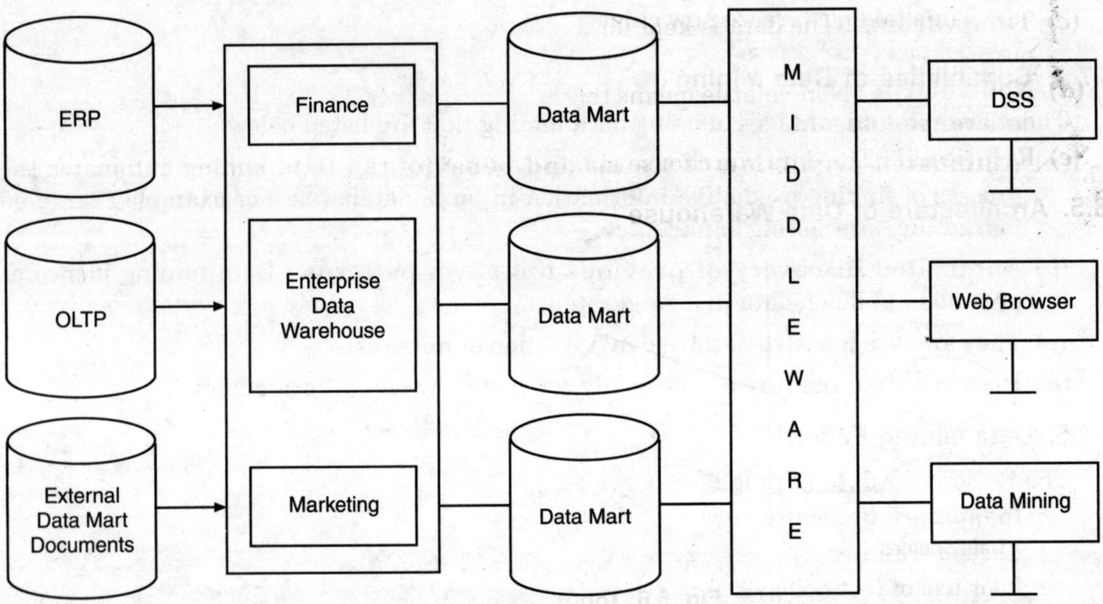

Fig. 5.4. Data Warehousing Framework

5.8.4. Characteristics of Data Warehousing

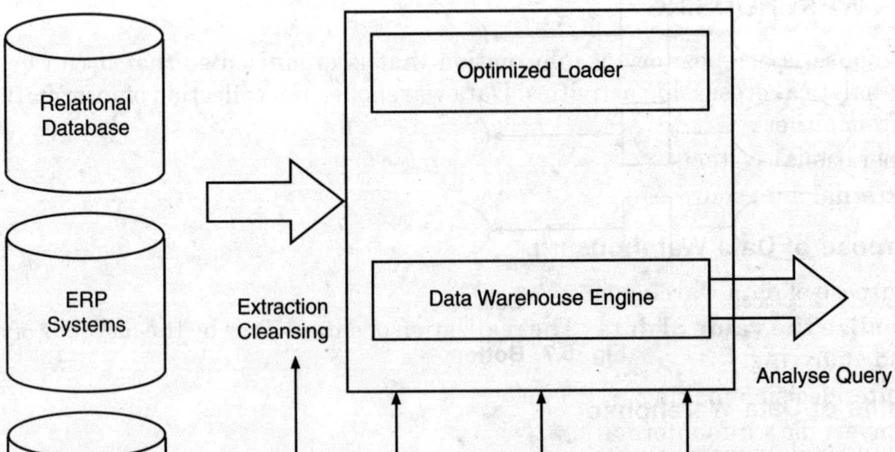

Fig. 5.5. Data Warehousing Functionality

The characteristics of data warehousing are :

(*a*) **Organisation :** The data is organized by subject like customer, vendor etc. It contains information relevant for making decision support only.

(*b*) **Consistency :** Consistency means that the data in different database may be encoded differently.

(*c*) **Time variant :** The data is kept for a long time so that it can be used for forecasting purpose.

(*d*) **Non-volatile :** Non-volatile means that the data cannot be updated when it enters the data warehouse.

(*e*) **Relational :** The data warehouse makes use of a relational database.

5.8.5. Architecture of Data Warehouse

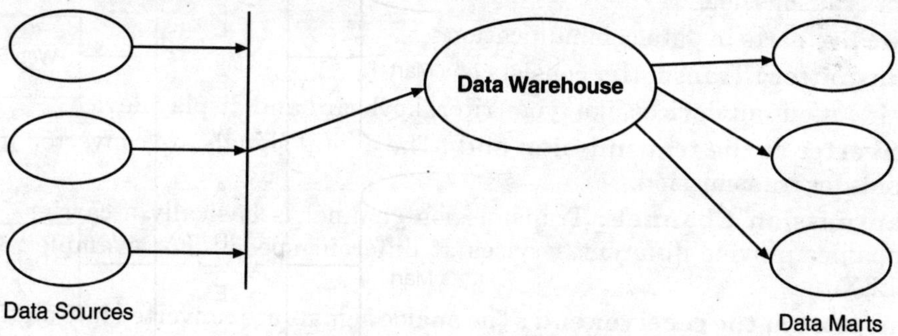

Fig. 5.6. Top-Down Model

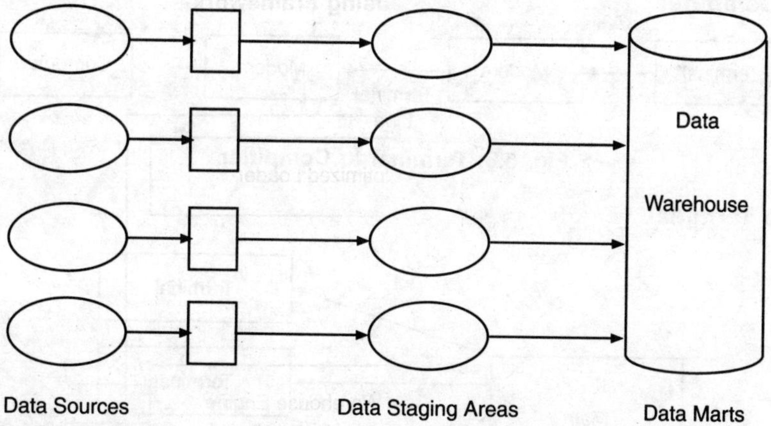

Fig. 5.7. Bottom-Up Model

5.8.6. Benefits of Data Warehouse

The benefits of data warehouse are :

- It has an ability to reach data quickly.
- It has an ability to reach data easily.
- It has an ability to reach data frequently.
- It helps in decision making process.
- It increases customer service.

- It increases customer satisfaction.
- It helps in restructuring business processes.

5.8.7. Problems in Data Warehouse

The problems faced in data warehouse are :

- A large amount of time is spent upon the extraction, cleaning and loading of data.
- Maintenance is required.

5.9. DATA COMMUNICATION

Data communication is an important part of computer utilization. It is divided into two parts :

(*a*) Data Processing

(*b*) Data Transmission

There are five parts in data communication :

(*a*) **Transmitter :** Transmitter consists of :

- Input-output devices like typewriter, keyboard and display device.

(*b*) **Converter at the transmission end :** The digital signals are converted into analog signals for transmission.

(*c*) **Transmission Channel :** Transmission channel is basically a carrier. Telephone companies provide different services at different speeds. For example : COMSAT, TELEX.

(*d*) **Converter at the receiver end :** The analog signals are converted into digital signals.

(*e*) **Receiver :** The receiver end consists of computer and a variety of input and output.

I. Terminal to Computer

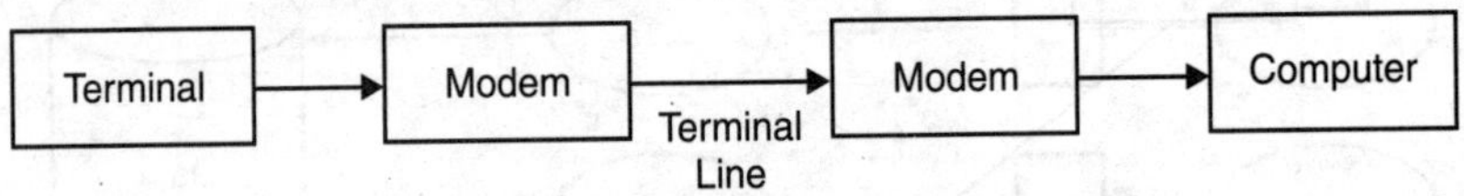

Fig. 5.8. Terminal to Computer

II. Computer to Terminal

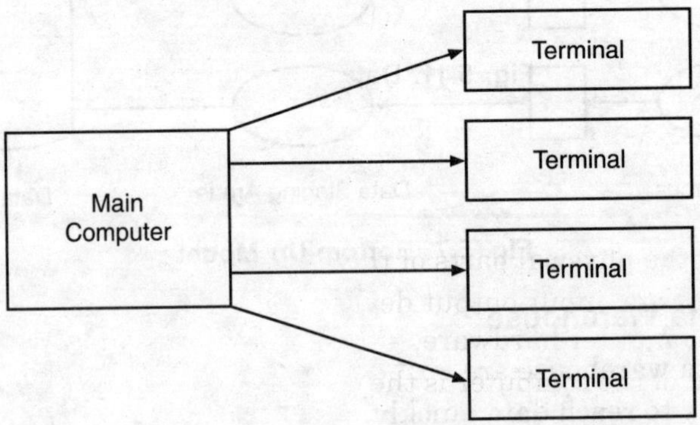

Fig. 5.9. Computer to Terminal

III. Data Communication Among Main Computer and Other Computers

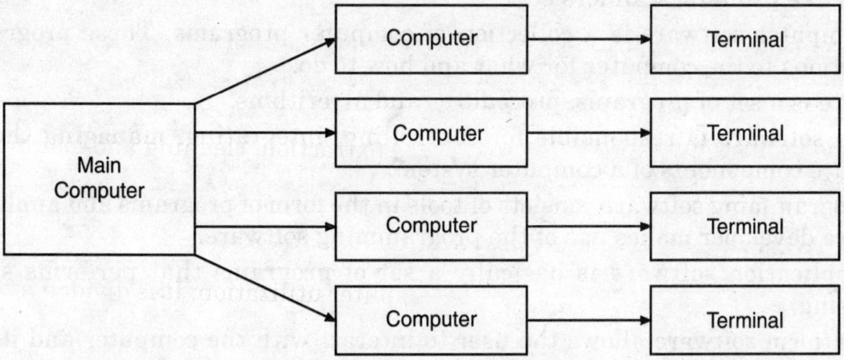

Fig. 5.10. Data Communication

IV. Data communication Among Several Main Computer and Their Terminals

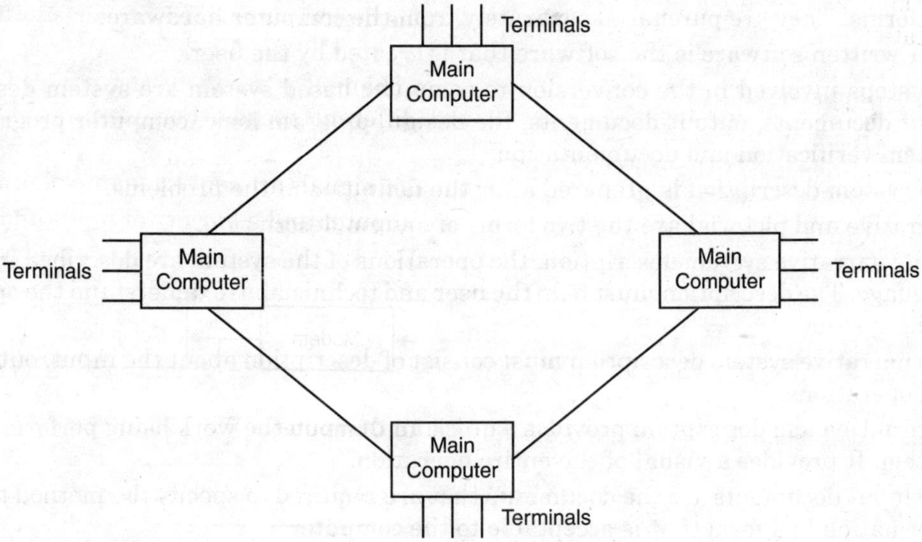

Fig. 5.11. Data Communication

Summary

- The hardware is the physical parts of the computer.
- Motherboard, storage, input-output devices, power supply and interface controller is the classification of computer hardware.
- The motherboard of the computer is the central circuit board that makes up the computer system.
- Storage of a computer system means those parts of the computer system that stores data for some period of time.
- The input-output device refers to the various interfaces that different functional units of system use to communicate with each other.

- Interface controller like USB, firewire is used to connect the computer to external peripheral devices like printers, scanners etc.
- The computer software is a collection of computer programs. These programs provide instructions to the computer for what and how to do.
- Software is a set of programs, procedures and algorithms.
- System software is responsible for controlling, integrating, managing the individual hardware components of a computer system.
- The programming software consists of tools in the form of programs and applications. The software developer makes use of the programming software.
- The application software is basically a set of programs that performs specific data processing.
- The platform software allows the user to interact with the computer and its peripheral devices.
- The computer software architecture consists of three layers. These are platform software, application software and user written software.
- Application software are the independent programs that are modified according to the platforms. They are purchased separately from the computer hardware.
- User written software is the software that is created by the user.
- The steps involved in the conversion to computer based system are system description, input documents, output documents, file design, program logic, computer programming, system verification and documentation.
- The system description is prepared after the definition of the problems.
- Narrative and pictorial are the two forms of system description.
- In the narrative system description, the operations of the system are described in English language. The description must help the user and technicians to understand the operations of the system.
- The narrative system description must consist of description about the input, output, files and operations.
- Pictorial system description provides a quick analysis on the work being performed by the system. It provides a visual of the entire operation.
- The input documents are the documents that are required to specify the method to put the information in a form that is acceptable to the computer.
- The output documents show the purpose of the entire operations that were performed on the input data. The management is concerned with the output data.
- The file design consists of the logic that is required to control the flow of data through the system.
- The program logic is a step by step representation of the method to complete the task.
- The computer programming is concerned with the method to instruct the processor on performing the operations on the input data so that the output is produced in the required format.
- In the system verification when the program is done with the compilation zone, it is placed in the memory in a machine readable format.
- Input, storage, central processor and output are the components of a computer system.
- Input is the data that is required to enter into the computer system. The input data should be in machine readable format.

- The central processor is the main part of the computer. It consists of three parts i.e. control section, arithmetic logic section and control input-output devices.
- The control section is responsible for integrating all the operations of the computer system. It basically directs and coordinates the operations.
- The arithmetic logic section performs the arithmetic and logic operations.
- The arithmetic operation is the calculation process and shifting of numbers.
- The logic operation is the decision to change the sequence of execution of the instructions.
- The data should be placed in the storage devices before performing the processing functions on the data.
- Storage is of two type's i.e. internal and external storage.
- Internal storage is known as memory and allows the machine to store data in electronic form.
- Direct access and sequential are the two types of external storage.
- Tne sequential storage requires that the data must be read from the beginning of the file so that the desired record is read and written.
- A database is a collection of data records, files and objects.
- A data base management system is a software package that creates the database, maintains the database and use the database. It can be accessed by multiple users and multiple application programmers at the same time.
- DBMS engine, data definition subsystem, data manipulation subsystem, application generation subsystem and data administration subsystem are the components of database.
- DBMS provides the facility of locking the data to avoid the problem of two users locking the data at the same time.
- DBMS makes sure that the data is accessed by authorized users. It is by default that the data of a user cannot be accessed by another user until the user gives permission of accessing the data.
- DBMS provides methods to maintain the integrity of the data. There are three integrity rules : domain, entity and referential.
- Hierarchical, network, relational, multi-dimensional and object-oriented are the five types of database models.
- The hierarchical database model is a model in which the data is organized in a tree like form.
- The network structure is a database model of representing objects and their relationships.
- Relational structure is a database model that makes use of two dimensional rows and columns to store the data.
- The object oriented structure is popular for the multimedia. It has an ability to handle graphics, pictures, voice and text.
- Managers are dependent on the database for retrieving and processing the data so that useful information can be produced.
- A relational database management system is a database management system that is based on a relational model.
- Data mining is a tool for analysing large amount of data that is present in a data warehouse. It helps in decision making process.
- Case based reasoning, neural computing and intelligent agents are the data mining techniques.

- Neural computing is a machine based learning approach. The historical data is examined for pattern recognition.
- Classification, clustering, association, sequencing and forecasting are the five types of information.
- Association identifies relationship between events that occur at one time.
- Forecasting eliminates the values that are based upon patterns within large sets of data.
- Retailing, sales, banking, manufacturing, production, insurance, computer hardware and software are the applications of data mining.
- Data warehouse is a repository of information that is organized so that it can be accessed easily for analytical processing activities.
- Staging area is a repository where the transaction data can be transferred for and in the data warehouse.
- Consistency means that data in different database may be encoded differently.
- Non-volatile means that the data cannot be updated when it enters the data warehouse.

Short Question Answers

1. What is computer hardware? Also classify it.

Ans. The hardware is the physical parts of the computer. Motherboard, storage, input-output devices, power supply and interface controller is the classification of computer hardware.

2. Classify the computer hardware in detail.

Ans. The computer hardware can be classified as :

(*a*) **Motherboard :** The motherboard of the computer is the central circuit board that makes up the computer system. It holds up CPU, main memory and slots for expansion cards.

(*b*) **Storage :** Storage of a computer system means those parts of the computer system that stores data for some period of time.

(*c*) **Input-Output devices :** The input-output device refers to the various interfaces that different functional units of system use to communicate with each other.

(*d*) **Power supply :** It is a case that holds transformer, voltage control and fan.

(*e*) **Interface controller :** Interface controller like USB, firewire is used to connect the computer to external peripheral devices like printers, scanners etc.

3. Explain computer software.

Ans. The computer software is a collection of computer programs. These programs provide instructions to the computer for what and how to do. Thus, software is a set of programs, procedures and algorithms.

4. Explain system software.

Ans. System software is responsible for controlling, integrating and managing the individual hardware components of a computer system. The users are not concerned with the low level details like transferring of data from the memory to disk.

5. What are the components of system software?

Ans. The components of system software are :

- Device drivers
- Operating systems
- Servers
- Window systems

- Management tools
- Disk formatters
- Networking control software
- File managers

6. What are the functions of system software?

Ans. The functions performed by the system software are :

- It is responsible for managing independent hardware components.
- Helps the application software programmer from the complex details of the system.
- Partition the computer resources in a safe and stable manner

7. What are the types of software?

Ans. The classification of software is done in two parts :

(*a*) **Programming Software :** The programming software consists of tools in the form of programs and applications. The software developer makes use of the programming software to perform the following functions :

- Create the programs
- Debug the programs
- Maintain the programs
- Support other programs

The examples of programming software are compilers, debuggers, interpreters, linkers and text editors.

(*b*) **Application Software :** The application software is basically a set of programs that performs specific data processing. It consists of many softwares.

8. Explain computer software architecture.

Ans. The computer software architecture consists of three layers. These are platform software, application software and user written software.

(*a*) **Platform Software :** The platform software allows the user to interact with the computer and its peripheral devices. It consists of firmware, device drivers, operating system and graphical user interface.

(*b*) **Application Software :** Application software are the independent programs that are modified according to the platforms. They are purchased separately from the computer hardware. For Example : Microsoft office suite, video games.

(*c*) **User-written software :** User written software is the software that is created by the user. For Example : spreadsheet templates, word processor templates.

9. What are the steps involved in the conversion of manual to computer based system?

Ans. The steps involved in the conversion of manual to computer based system are system description, input documents, output documents, file design, program logic, computer programming, system verification and documentation.

10. Explain system description.

Ans. The system description is prepared after the definition of the problems. System description consists of statements related to the input of the system, output of the system, operations performed and files required. The purpose of system description is logical information flow and logical operations. These operations are used to carry out the particular system design.

11. What are the forms of System Description?

Ans. There are two forms of system description. These are:

1. **Narrative system description :** In the narrative system description, the operations of the system are described in English language. The description must help the user and technicians to understand the operations of the system. The narrative system description must contain description about the input, output, files and operations.
2. **Pictorial system description :** Pictorial system description provides a quick analysis on the work being performed by the system. It provides a visual of the entire operation. For example : System flowchart.

12. Explain input documents.

Ans. The input documents are the documents that are required to specify the method to put the information in a form that is acceptable to the computer. The computer acceptable form is the machine usable form. The input format is selected on the basis of volume of information, frequency, accuracy and verification.

13. Explain output documents.

Ans. The output documents show the purpose of the entire operations that were performed on the input data. The management is concerned with the output data.

14. Explain file design.

Ans. The file design consists of the logic that is required to control the flow of data through the system. It is a part of the system design only. It should be noted that the flow of data depends solely on the design of the data files.

15. Explain program logic.

Ans. In the program logic, program flowcharts are used. The program logic is a step by step representation of the method to complete the task. It deals with the entire task in a systematic manner. It is used to organize facts, outline problems, outline logics and outline solutions.

16. Explain computer programming.

Ans. The computer programming is concerned with the method to instruct the processor on performing the operations on the input data so that output is produced in the required format.

It consists of different programming languages like C, C++ and Pascal.

17. Explain system verification.

Ans. In the system verification when the program is done with the compilation zone, it is placed in the memory in a machine readable format. The program is ready to perform the operations like process the terminal input, update the master file on the disk and print the report.

18. Explain documentation process.

Ans. The documentation process is the most important concept. The documentation process is of three types.

(*a*) **Input :** The documentation should consist of simple description about the system. It should provide a clear description of the required inputs in the system.

(*b*) **Maintenance :** The documentation process should cover the technical documentation.

(*c*) **Output :** The documentation should consist of the simple description about the system. It should provide a clear description of the outputs and their limitations.

19. What are the components of a computer system?

Ans. Input, central processor, storage and output are the components of a computer system.

20. Explain control section.

Ans. The control section is responsible for integrating all the operations of the computer system. It basically directs and coordinates the operations.

21. Explain control of input-output devices.

Ans. The control of input and output devices is associated with two factors. First is the entry and removal of the information from the storage devices. Second is the routing of information between the storage devices and arithmetic logic section.

22. Explain arithmetic logic section.

Ans. The arithmetic logic section performs the arithmetic and logic operations. The arithmetic operation is the calculation process and shifting of numbers. The logic operation is the decision to change the sequence of execution of the instructions.

23. Explain internal storage.

Ans. Internal storage is known as memory. It allows the machine to store data in electronic form. It allows the machine to store long instructions that is required to guide the machines regarding what to do.

24. Explain external storage.

Ans. External storage is of two types, i.e. direct access and sequential access.

25. Explain direct access.

Ans. In the direct access, the data can be read from anywhere. It is not necessary that the required data should be read from the beginning of the file. For Example : disk and magnetic drums.

26. Explain sequential access.

Ans. The sequential storage requires that the data must be read from the beginning of the file so that the desired record is read and written. For Example : magnetic tapes.

27. What is DBMS?

Ans. A database is a collection of data records, files and objects. A data base management system is software package that creates the database, maintains the database and use the database. It can be accessed by multiple users and multiple application programmers at the same time.

28. What are the components of database?

Ans. The components of a database are DBMS engine, data definition subsystem, data manipulation subsystem, application generation subsystem and data administration subsystem.

29. Explain DBMS engine.

Ans. DBMS engine accepts requests from other database systems and converts them into physical parts. It accesses the database and data dictionary.

30. Explain data definition subsystem.

Ans. Data definition subsystem helps the user to create the data dictionary and define the structure of the files.

31. Explain data manipulation subsystem.

Ans. Data manipulation subsystem helps the user to manipulate the system. It helps the user to add, modify and delete the information from the database.

32. Explain application generation subsystem.

Ans. Application generation subsystem helps the user to develop applications. The user should perform a detailed series of tasks to process a transaction.

33. Explain data administration subsystem.

Ans. Data administration subsystem helps the user to manage the entire database by providing facilities like backup and recovery, security management etc.

34. Explain hierarchical model.

Ans. The hierarchical database model is a model in which the data is organized in a tree like form. The hierarchical model is a simple model. It represents information using one to many relationships. For example : Storing geographical information, mobile etc.

35. Explain network structure.

Ans. The network structure is a database model of representing objects and their relationships. It is a flexible model. The network structure implements many to many relationships.

36. Explain relational structure.

Ans. Relational structure is a database model that makes use of two dimensional rows and columns to store the data. It is a complex method and is not suitable for the end user. The relational structure is used by the mainframe, midrange and microcomputer systems.

37. Explain multidimensional structure.

Ans. The multidimensional structure is very similar to the relational model. The multidimensional structure is quite easy to maintain and understand. The records are stored as fundamental attributes.

38. Explain object-oriented structure.

Ans. The object oriented structure is popular for the multimedia. It has an ability to handle graphics, pictures, voice and text. It is designed to work with the object oriented programming languages like java.

39. Explain RDBMS.

Ans. A relational database management system is a database management system that is based on a relational model. It is designed by E.F. Codd. It is based on the theory of sets and relations.

40. Explain data mining.

Ans. Data mining is a tool for analysing large amount of data that is present in a data warehouse. It helps in decision making process. It is also known as "Knowledge discovery of database". The goal of the data mining process is to extract knowledge from the existing data and transform into a human understandable format so that it can be used further.

41. Explain data mining techniques.

Ans. The most common used techniques in data mining are :

(*a*) **Case-based reasoning :** This uses historical cases to recognize patterns.

(*b*) **Neural computing :** Neural computing is a machine based learning approach. The historical data is examined for pattern recognition. These patterns are used for decision support.

(*c*) **Intelligent agents :** It is an approach for retrieving information from the Internet. A large amount of information is available on the Internet.

(*d*) **Other Tools :** Many tools like decision trees, genetic algorithms are used in the data mining process.

42. Explain the features of database.

Ans. Data sharing, data concurrency, data security, data locking, data integrity, fault tolerance, fault recovery and language support are the features of database.

43. What are the advantages of using database?

Ans. The advantages of database are :

- It keeps a check on the data redundancy.
- It ensures data integrity.
- It ensures that the data can be shared across all applications.
- It contains a backup and recovery procedure of data.
- It provides better services to the users.
- Improved flexibility.
- Cost of developing and maintaining the system is low.
- Improved security.

44. What are the disadvantages of using database?

Ans. The disadvantages of using database are :

- Implementation of a database can be expensive.
- Implementation of a database can be time consuming.
- Even with the provision of high security, it is possible for unauthorized users to access the database.

45. What are the advantages of RDBMS?

Ans. The advantages of RDBMS are :

- It is simple in nature.
- It is easy to understand.
- It is easy to manipulate.
- It allows having a dynamic relationship between the entities.

46. Classify information types.

Ans. Classification, clustering, association, sequencing and forecasting are the information types.

47. What are the applications of data mining?

Ans. Retailing, sales, banking, manufacturing, production, insurance, computer hardware and software are the applications of data mining.

48. What are the advantages of data mining?

Ans. There are some advantages of using data mining that are listed below :

(*a*) Data mining automates the process of finding predictive information in large databases. For example : targeted marketing, forecasting bankruptcy.

(*b*) Data mining identifies previous unknown data in a single step.

(*c*) They analyse massive database in fraction of minutes.

(*d*) Data mining procedure can be easily conducted by non-programmers.

49. What are the reasons for data mining failure?

Ans. The reasons for data mining failure are :

- Inappropriate architecture
- Poor use of technology
- Lack of user input
- Lack of training
- Lack of data quality
- Interference of corporate politics

50. Explain data warehousing.

Ans. Data warehouse is a repository of information that is organized so that it can be accessed easily for analytical processing activities. Data warehouse is a collection of information that is obtained from operational systems and external data sources.

51. What is the purpose of data warehouse?

Ans. The purpose of data warehousing is :

- The realization of data is done by methods of reporting and analysing.
- Better decision making
- Convert data into information
- Create a competition

52. What are the components of data warehouse?

Ans. There are three components of data warehousing. These are :

(*a*) **Staging Area :** Staging area is a repository where the transaction data can be transferred for and in the data warehouse.

(*b*) **Data Mart :** The data warehouse is the union of a set of data marts.

(*c*) **Operational Data Store :** The operational data store is modelled to support the real time reporting needs.

53. What are the characteristics of data warehouse?

Ans. The characteristics of data warehousing are :

(*a*) **Organisation :** The data is organized by subject like customer, vendor etc. It contains information relevant for making decision support only.

(*b*) **Consistency :** Consistency means that data in different database may be encoded differently.

(*c*) **Time variant :** The data is kept for a long time so that it can be used for forecasting purpose.

(*d*) **Non-volatile :** Non-volatile means that the data cannot be updated when it enters the data warehouse.

(*e*) **Relational :** The data warehouse makes use of a relational database.

54. What are the benefits of data warehouse?

Ans. The benefits of data warehouse are :

- It has an ability to reach data quickly.
- It has an ability to reach data easily.
- It has an ability to reach data frequently.
- It helps in decision making process.
- It increases customer service.
- It increases customer satisfaction.
- It helps in restructuring business processes.

55. Explain data communication.

Ans. Data communication is an important part of computer utilization. It is divided into two parts *i.e.*, data processing and data transmission.

Review Questions

1. Explain computer hardware and software in detail.
2. Explain conversion of manual to computer based system. Also explain the components of a computer system.
3. Explain DBMS and RDBMS.
4. Explain data mining and data warehousing.
5. Explain data communication in detail.

Section-B

Inside This Section :

Chapter 6

Conceptual System Design

"That growth combined with changing requirements in areas like governance and protection, has resulted in increased costs and information management challenges."

Inside this Chapter

6.1. CONCEPTUAL SYSTEM DESIGN OVERVIEW

The conceptual system design is an extension of the logical design of the system. The conceptual system design takes into account the various technological factors that are required for achieving the functional requirements of the system. It is also known as feasibility design, gross design or high level design.

The **input** of the Conceptual System Design is :

- Management information requirements
- Management objectives

The **output** of the Conceptual System Design is :

- Set of documents that describe the MIS in full detail. It helps the technicians to start their work on the detailed design of the system.

The **key tasks** that are performed in the **Conceptual Designing** of the system are :

- Defining the problems in a complete manner.
- Refining the management objectives to set the system objectives.
- Establishing the system constraints.
- Determining the Information needs and requirements.
- Developing the alternatives design and selecting an appropriate one.
- Documenting the conceptual system design phase.

6.2. DEFINE THE PROBLEMS

The first appropriate step that must be taken by the management of an organization is the defining of the problems. These are the problems that are to be solved by an organization. It must be noted that the objectives of an organization are derived with the help of mission statement for a business organization. From the objectives, plans are derived. With each objective and plan, information needs for an organization is derived.

The process of defining the problems is :

(*a*) State the information needs of an organization.

(*b*) Ask questions related to those stated needs.

(*c*) Suggest explanations of that need.

(*d*) Detail the original statement.

(*e*) Review the detailed original statement.

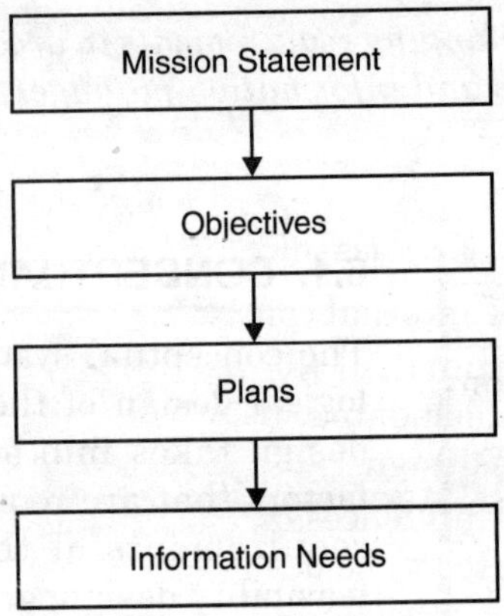

Fig. 6.1. Problem Definition

6.3. SET SYSTEM OBJECTIVES

The system objectives of an organization must be set. System objectives are expressed in terms of what the manager can do when the information requirements of an organization is fulfilled.

- The setting of system objectives of an organization that cut across all the functional areas is very difficult.
- It is easy to determine the objectives of an operational system that is served by a main system.

The system objectives are viewed by the following people :

(*a*) **Technician :** The technician sees the system objectives of the smaller system as an input to a larger system.

(*b*) **Manager :** It is the responsibility of the manager to define the system objectives in terms of the information needs. They should not define the terms according to the satisfaction of demands.

The system objectives must be stated in terms of :

- Objectives of the department
- Objectives of the group
- Objectives of the function
- Manager to be served

6.4. ESTABLISH SYSTEM CONSTRAINTS

Before understanding the system constraints, first understand the meaning of the word **"Constraints"**. Constraints are the problem boundaries or limitations.

A constraint helps the designer to modify the conditions under which the objectives can be fulfilled. It also helps the designer to consider the limitations that restrict the design. A constraint can have both positive and negative effect.

But now the question is what the advantage of setting up a constraint is. The answer is very simple. Constraint helps us to ensure that a design is realistic.

6.4.1. Classification of Constraints

Constraints can be classified as internal and external constraints.

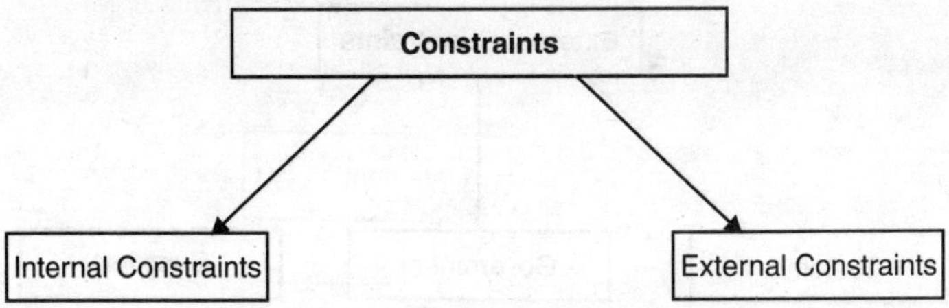

Fig. 6.2. Classification of Constraints

Internal Constraints : The internal constraints are of five types. These are :

(*a*) **Top Management Support :** If the top management support is not obtained, then the design cannot be implemented. It is the responsibility of the top management to approve and support the design.

This internal constraint influences on the decision of the management on the kind of the system that is supposed to be designed.

(*b*) **Policy Considerations :** Policy considerations should be kept in mind. It helps in setting the limit for objectives. It also helps in modifying an already existing approach to design a system. The flow of information is influenced by the :

- Structure of the organization
- Position of the managers

The organizational policies limit the approach to system design.

(*c*) **Personnel Needs :** The personnel needs are a major limiting factor in the :

- Design of Information System
- Utilization of Information System

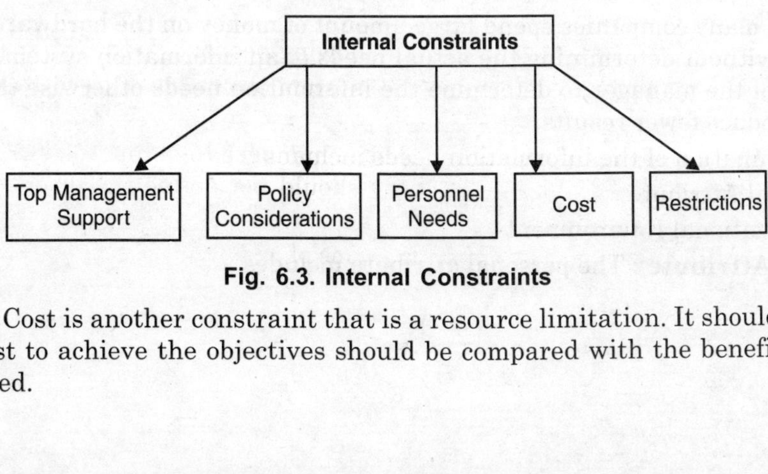

Fig. 6.3. Internal Constraints

(*d*) **Cost :** Cost is another constraint that is a resource limitation. It should be noted that the cost to achieve the objectives should be compared with the benefits that will be achieved.

(e) Restrictions : Restriction is the constraint that is made by the managers and designers on the design of an information system. In the designing phase, the manager needs to alter some of the objectives so that the system is fit with other outputs and equipment. The manager will also restrict on the time and effort that is applied to the entire phase.

External Constraints : The external constraints are of three types. These are :

(a) Customer : Some of the systems like :

- Order entry
- Order billing

Should be designed keeping in mind the needs of the customer. If some outputs from the system are not acceptable, customer constraint may be faced.

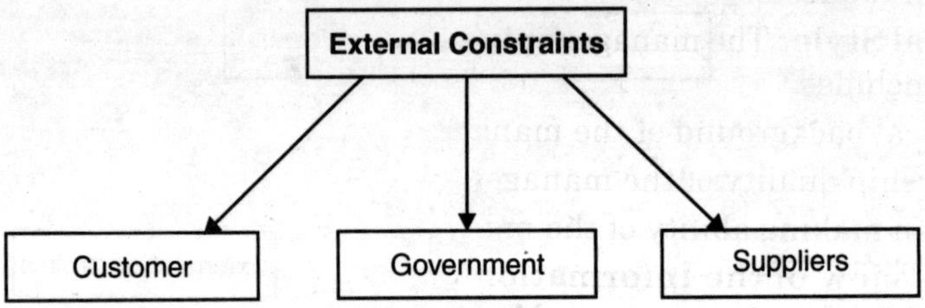

Fig. 6.4. External Constraints

(b) Government : The governmental constraint is of three types :

- Federal
- State
- Local

This type of constraint imposes restrictions on :

- Maintenance of the security of information
- Meeting of certain procedures of record keeping
- Reporting to stockholders

(c) Suppliers : The suppliers must also be taken into account during the designing process of an information system.

6.5. DETERMINE INFORMATION NEEDS

The information needs are very necessary in designing the information system. There should be a clear statement of the information need. Thus the information need is the fundamental and necessary requirement in designing an information system.

Nowadays, many companies spend large amount of money on the hardware and software of the system without determining the actual needs of an information system. It is thus the responsibility of the manager to determine the information needs otherwise the information system will produce fewer results.

The determination of the information needs includes :

1. Personal Attribute
2. Organizational Environment

Personal Attribute : The personal attribute includes :

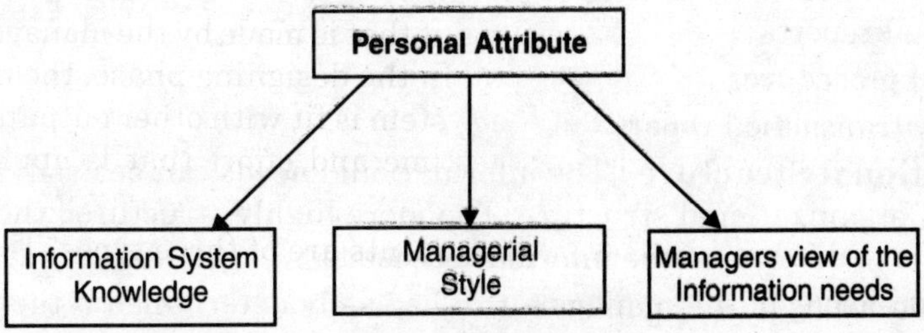

Fig. 6.5. Personal Attribute

(a) Information System Knowledge : If the managers are aware about the working and operations of the information system, it helps in making more appropriate information requests.

(b) Managerial Style : The manager's style affects the kind of information that is required. The style includes :

- Technical background of the manager
- Leadership quality of the manager
- Decision making ability of the manager

(c) Managers view of the Information needs : The view of the information system is different for different managers. Most of the managers do not pay attention on the type of information that is required.

Organizational Environment : The organizational environment includes :

(a) Nature of the Company : It is the nature of the company that decides functions like :

- Communication
- Operations

These functions depend upon the company's size and complexity. It must be noted that the larger the firm is more formal information systems are required.

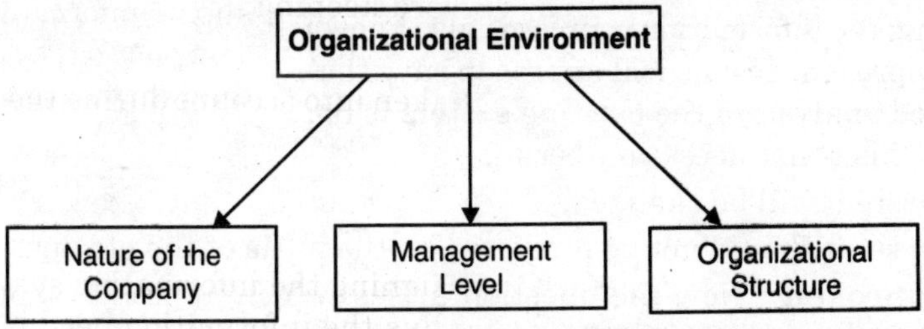

Fig. 6.6. Organizational Environment

(b) Management Level : The management level includes three levels. These are :

- **Strategic Planning :** The strategic planning includes :
  - One time report
  - Inquiry
  - Summary
- **Management Control :** The management control includes :
  - Exception report
  - Different regular reports
  - Summary
- **Operational Control :** The operational control includes :

- Formal reports
- Fixed procedures
- Daily transaction report

(c) Organizational Structure : The informational needs can be easily determined with the help of organizational structure. The more highly structured the organization is, the easier is to determine the information needs.

The information needs of the managers can be easily determined if the following factors are always considered :

- Understanding the relationships
- Defining decision making areas properly

6.6. DETERMINE INFORMATION SOURCES

Determination of the information needs and information sources are quite corresponding. The natural way of finding information is through :

- Books
- Records
- Files

The determination of information sources consists of two theories :

(*a*) **Redesigning of the new system from the old system :** If a small modification is made in the existing system, then the information requirements can be fulfilled. This helps in reducing the designing effort. A study should be done to know the specific areas that need improvement. A study should be done for knowing the :

- Common input sources
- Data linked with the new design
- Cost associated with the new design

(*b*) **Logical Approach :** The logical approach of determining the information sources is also known as fresh approach. The logical approach says that the detailed analysis of the existing system is not required. This is not necessary because :

- New system will be changed.
- New system should not be predicted on the basis of the old one.

> **DO YOU KNOW?** ✍
>
> The vision is really about empowering workers, giving them all the information about what's going on so they can do a lot more than they have done in the past.
>
> **–Bill Gates**

Choice of Approach : Now the question arises, which approach should be used? The choice of approach is done :

- It is done either by the manager or designer.
- It is done on the basis of the information sources.
- It is done on the basis of the objectives of the system that is being designed.
- It is done on the preference of the old or new system.

Analysis and integration of the Information Sources : The source of information is broadly classified into :

(a) Internal and External Records : The records are of two types. One is internal and another is external record.

Internal Record : Internal records are of written type. These documents contain information about the existing system. For example :

- File record

- Letters
- Reports
- Documents

External Record : External record is the record that comes from various different sources. For example :

- Government Statistics

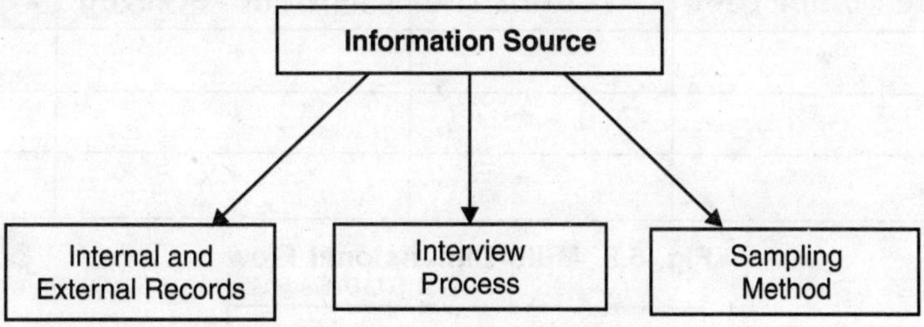

Fig. 6.7. Information Source

(*b*) Interview Process : Interview process of managers and personnel's is a method of identifying the possible sources of information. It helps in analyzing the existing information system. The interview process helps in obtaining secure information. The **problem** faced while conducting interviews is human made errors. The **solution** to overcome this error is through :

- Proper planning
- Increasing confidence level

(*c*) Sampling Method : Sampling method is performed when there is a large amount of data available. In this method, only a small part of data is examined. The advantage of sampling method is :

- Less time is required to perform the sampling operation.
- Less cost is faced to perform the operations.

For Example : Work Sampling

The work sampling method is used to perform the following operations :

- Action of people
- Action of machine

These operations are sampled in terms of time.

Analysis and Synthesis : Analysis and synthesis makes use of two techniques to avoid duplication. These techniques are :

(*a*) Input-Output analysis : The analysis on input and output is done on the basis of :

- Input-output chart
- Visual display of the information inputs
- Information output

	Paychecks	**Payroll**	**Governmental**	
Name	–		–	
Start time		–	–	
Stop time		–	–	
Inputs		–	–	**Outputs**

Fig. 6.8. Input-Output analysis

Note : The dot helps in determining the relationship.

(*b*) **Multi-Dimensional Flow :** Multi-dimensional flow is a technique of organizing the various information sources. This technique makes use of the flowchart. The flowchart helps in tracing the flow of information from source to destination. The multi-dimensional flowchart should be arranged in an increasing order.

	Customer file	Vendor file	Payment received	Payment paid
Customer	*		*	
Marketing		*	*	
Production		*	*	

Fig. 6.9. Multi-Dimensional Flow

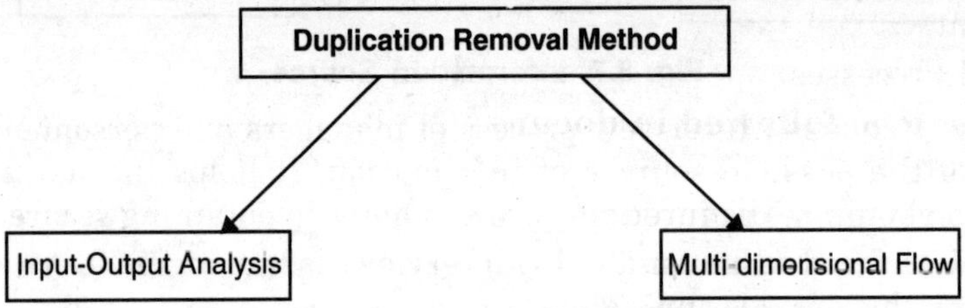

Fig. 6.10. Duplication Removal Method

6.7. DEVELOP ALTERNATIVE CONCEPTUAL DESIGN

Alternative conceptual designs should be made. We must briefly note the various constituents of the conceptual design. These are :

- Decision points
- Information flow patterns
- Information channel
- Manager's role
- Relationship between the operating and management information system.

The alternative designs should be measured in terms of its advantages and disadvantages.

Evaluation method : The evaluation method is a method to compare alternative designs. The method consists of the following criterion :

- The objectives of the system must be compared with the expected performance of the design of the system.
- Prepare a cost effective analysis of the system.
- Flowcharts must be examined. It helps to identify the strong and weak points of the system design.
- Study the design in more detail.
- Study the number of operations performed by the system and its potential.
- Examine the database quality.

6.8. DOCUMENT THE SYSTEM CONCEPT

Now the major factor is the documentation of the system concept. Documentation is required because large information has been collected. It helps in making a more detailed description of the system.

> **DO YOU KNOW?**
>
> Communication can't always follow the top-down model. With the fluidity of information in business today, leaders need to be masterful listeners, they need an ability to receive as well as send.
>
> **–Joseph Badaracco**

Documentation should be done of :

- Flowchart
- Input
- Output
- Operations
- Flow of information

Role of Manager : Manager is also involved in the documentation of the system. The manager is concerned with the amount to which the system provides the required information.

Role of Designer : Designer is also involved in the documentation of the system. The designer is related to :

- Materials
- Equipment
- Technical Processes

Instructions to be followed in documentation process : The following instructions should be taken care of :

- What data should be captured?
- When should the files be used?
- How the processing should be done?
- What output will be generated?
- How the files will be distributed?
- How the output should be distributed?

Black Box Concept : The black box concept includes the following elements :

- Input
- Output
- Master files
- Data processing rules

But why it is known as black box concept? The answer is quite simple. It is known as black box because the processing element remains black to the manager and user.

There are certain rules that should be followed in the black box concept. These are :

- Processing of the input data against the data file and finally produce an output.
- Processing the input data for updating the file.
- Processing of input data to output data without referencing the data file.
- Producing an output from the file without processing the input.

Generate System Flow : The generation of system flow is a method of specifying the general structure of a computer based information system. It reflects the design effort.

Characteristics : The characteristics of the system flow are :

- System objectives are easily achieved.
- The system objectives are reflected in the flow diagram.
- Information needs are designed into the system.
- Information sources are designed into the system.
- Subsystems are integrated into one system.
- Decision rules and points are shown.

System Inputs : The system inputs are the ones that are received from outside the firm and are converted to machine usable form. The documentation of system inputs must be done in a detailed manner. The documentation must include the following :

- Source of the input data
- What is the form of the input data?
- Who is responsible for the production of the input data?

System designers are concerned with the system inputs. System inputs are collected with the help of forms. Thus it is the responsibility of the designer to design the form in an appropriate manner.

System Outputs : The system output is the one that is received from the processing of the input data. The documentation must include the following :

- Where does the system output go?
- What form they take?
- Who is responsible for receiving them?
- How many copies of the output form are required?
- What kind of output form is required?

Documentation of Inventory Management : The documentation of the inventory management of the raw materials is concerned with the raw materials that are required for the manufacturing process. The inventory management of raw material includes four plants. Each plant has its own computers. The three subsystems present are :

- Stock level subsystem
- Reorder point subsystem
- Economic order quantity

Documentation of Stock Level : In the documentation of stock level, as soon as the material is received, the receiving report is marked to show :

- Quantity acceptance
- Quality acceptance

There is a matching done between the received report and receiving order about the :

- Quantity acceptance
- Quality acceptance
- Data acceptance

Finally the receiving order is sent to the computer for updating the inventory file.

6.9. PREPARE CONCEPTUAL DESIGN REPORT

A conceptual design report is prepared to know the amount spend in the organizational changes. The design report must consist of the following parameters :

- Objectives
- Summary
- Nature of the system
- Reason of selecting alternative design
- Time required to design the system
- Time required to implement the system
- Resources required to design the system
- Resource required to implement the system
- Performance specifications

Summary

- The conceptual system design is an extension of the logical design of the system.
- The conceptual system design takes into account the various technological factors that are required for achieving the functional requirements of the system.
- The **input** of the Conceptual System Design is management information requirements and objectives.
- The output of the Conceptual System Design is the set of documents that describe the MIS in full detail. It helps the technicians to start their work on the detailed design of the system.
- The first appropriate step that must be taken by the management of an organization is the defining of the problems. These are the problems that are to be solved by an organization.
- System objectives are expressed in terms of what the manager can do when the information requirements of an organization is fulfilled.
- The technician sees the system objectives of the smaller system as an input to a larger system.
- The manager must define the system objectives in terms of the information needs. They should not define the terms according to the satisfaction of demands.
- A constraint helps the designer to modify the conditions under which the objectives can be fulfilled. It also helps the designer to consider the limitations that restrict the design.
- Internal and external constraints are the two types of constraints.
- Top management support, policy considerations, personal needs, cost and restrictions are the various types of internal constraints.
- It is the responsibility of the top management to approve and support the design. Top management internal constraint influences the decision of the management on the kind of the system that is supposed to be designed.
- A policy consideration helps in setting the limit for objectives. It also helps in modifying an already existing approach to design a system.
- The personnel needs are a major limiting factor in the design and utilization of an Information System.
- Restriction is the constraint that is made by the managers and designers on the design of an information system.
- Customer, government and suppliers are the three types of external constraints.
- The governmental constraint is of three types i.e. federal, state and local.
- The information needs are very necessary in designing an information system. There should be a clear statement of the information need. Thus the information need is the fundamental and necessary requirement in designing an information system.
- The determination of the information needs includes personal attribute and organizational environment.
- The personal attribute includes information system knowledge, managerial style and managers view of the information needs.
- The organizational environment includes nature of the company, management level and organizational structure.
- The management level includes three levels i.e. strategic planning, management control and operational control.
- The strategic planning includes one-time report, inquiry and summary.

- The management control includes exception report, different regular reports and summary.
- The operational control includes formal reports, fixed procedures and daily transaction report.
- The informational needs can be easily determined with the help of organizational structure. The more highly structured the organization is, the easier is to determine the information needs.
- Determination of the information needs and information sources are quite corresponding. The natural way of finding information is through books, records and files.
- The determination of information sources consists of two theories i.e. logical approach and redesigning of the new system from the old system.
- The logical approach of determining the information sources is also known as fresh approach. The logical approach says that the detailed analysis of the existing system is not required.
- The records are of two types. One is internal and another is external record.
- Internal records are of written type. These documents contain information about the existing system.
- External record is the record that comes from various different sources.
- Information sources are internal and external record, interview process and sampling method.
- Interview process of managers and personnel's is a method of identifying the possible sources of information. It helps in analyzing the existing information system. The interview process helps in obtaining secure information.
- Sampling method is performed when large amount of data is available. In this method, only a small part of data is examined.
- Analysis and synthesis makes use of two techniques to avoid duplication. These techniques are input-output analysis and multi-dimensional flow.
- The evaluation method is a method to compare the alternative designs.
- Documentation is required because large information has been collected. It helps in making a more detailed description of the system.
- Manager is also involved in the documentation of the system. The manager is concerned with the amount to which the system provides the required information.
- Designer is also involved in the documentation of the system. The designer is related to materials, equipment and technical processes.
- In the black box concept, the processing element remains black to the manager and user.
- The generation of system flow is a method of specifying the general structure of a computer based information system. It reflects the design effort.
- The system inputs are the ones that are received from outside the firm and are converted to machine usable form. The documentation of system inputs must be done in a detailed manner.
- The system output is the one that is received from the processing of the input data.
- The documentation of the inventory management of the raw materials is concerned with the raw materials that are required for the manufacturing process. The inventory management of raw materials includes four plants. Each plant has its own computers.
- The three subsystems present in the inventory management are stock level subsystem, reorder point subsystem and economic order quantity.
- A conceptual design report is prepared to know the amount spend in the organizational changes.

Short Question Answers

1. Give a conceptual design overview.

Ans. The conceptual system design is an extension of the logical design of the system. The conceptual system design takes into account the various technological factors that are required for achieving the functional requirements of the system. It is also known as feasibility design, gross design or high level design.

2. What is the input of the conceptual system design?

Ans. The input of the conceptual system design is management information requirements and management objectives.

3. What is the output of the conceptual system design?

Ans. The output of the conceptual system design is the documents that describe the MIS in full detail. It helps the technicians to start their work on the detailed design of the system.

4. What are the key tasks performed in the conceptual system design?

Ans. The key tasks that are performed in the conceptual designing of the system are :

- Defining the problems in a complete manner.
- Refining the management objectives to set the system objectives.
- Establishing the system constraints.
- Determining the information needs and requirements.
- Developing the alternatives design and selecting an appropriate one.
- Documenting the conceptual system design phase.

5. Explain the concept of defining the problems.

Ans. The first appropriate step that must be taken by the management of an organization is the defining of the problems. These are the problems that are to be solved by an organization. The objectives of an organization are derived with the help of mission statement for a business organization. From the objectives, plans are derived. With each objective and plan, information needs for an organization is derived.

6. Explain the process of defining the problems.

Ans. The process of defining the problems is :

(*a*) State the information needs of an organization.

(*b*) Ask questions related to those stated needs.

(*c*) Suggest explanations of that need.

(*d*) Detail the original statement.

(*e*) Review the detailed original statement.

7. Explain the process of setting the system objectives.

Ans. The system objectives of an organization must be set. System objectives are expressed in terms of what the manager can do when the information requirements of an organization is fulfilled.

8. Who are the people that view the system objectives?

Ans. The system objectives are viewed by the technician and manager. The technician sees the system objectives of the smaller system as an input to a larger system. It is the responsibility of the manager to define the system objectives in terms of the information needs. They should not define the terms according to the satisfaction of demands.

9. What are the terms in which the system objectives must be stated?

Ans. The system objectives must be stated in terms of :

- Objectives of the department
- Objectives of the group

- Objectives of the function
- Manager to be served

10. Define constraint.

Ans. A constraint helps the designer to modify the conditions under which the objectives can be fulfilled. It also helps the designer to consider the limitations that restrict the design. A constraint can have both positive and negative effect.

11. What are the two types of constraints?

Ans. Internal and external are the two types of constraints.

12. What are the five types of internal constraints?

Ans. Top management support, policy considerations, personal needs, cost and restrictions are the various types of internal constraints.

13. Explain the top management support internal constraint.

Ans. If the top management support is not obtained, then the design cannot be implemented. It is the responsibility of the top management to approve and support the design. Top management internal constraint influences on the decision of the management on the kind of the system that is supposed to be designed.

14. Explain policy consideration internal constraint.

Ans. Policy considerations should be kept in mind. It helps in setting the limit for objectives. It also helps in modifying an already existing approach to design a system. The organizational policies limit the approach to system design.

15. Explain personal need internal constraint.

Ans. The personnel needs are a major limiting factor in the design and utilization of Information System.

16. Explain restriction internal constraint.

Ans. Restriction is the constraint that is made by the managers and designers on the design of an information system. The manager needs to alter some of the objectives so that the system is fit with other outputs and equipment. The manager will also restrict on the time and effort that is applied to the entire phase.

17. What are the three types of external constraint?

Ans. Customer, government and suppliers are the three types of external constraints.

18. Explain customer constraint.

Ans. Systems like order entry and billing should be designed keeping in mind the needs of the customer. If some outputs from the system are not acceptable, customer constraint may be faced.

19. What are the three types of governmental constraint?

Ans. The governmental constraint is of three types i.e. Federal, state and local.

20. Explain the supplier constraint.

Ans. The suppliers must also be taken into account during the designing process of an information system.

21. Explain determination of information needs.

Ans. The information needs are very necessary in designing the information system. There should be a clear statement of the information need. Thus the information need is the fundamental and necessary requirement in designing an information system.

22. What are the two parameters included in the determination of information needs.

Ans. The determination of the information needs includes personal attribute and organizational environment.

23. Explain personal attribute.

Ans. The personal attribute includes information system knowledge, managerial style and managers' view of the information needs.

24. Explain organizational environment.

Ans. The organizational environment includes nature of the company, management level and organizational structure.

25. What are the three components of the management level?

Ans. The management level includes three levels i.e. strategic planning, management control and operational control.

26. Explain managerial style.

Ans. The manager's style affects the kind of information that is required. The style includes technical background of the manager, leadership quality of the manager and decision making ability of the manager.

27. What are the components of strategic planning?

Ans. The strategic planning includes one-time report, inquiry and summary.

28. What are the components of management control?

Ans. The management control includes exception report, different regular reports and summary.

29. What are the components of operational control?

Ans. The operational control includes formal reports, fixed procedures and daily transaction report.

30. Explain determination of information sources.

Ans. Determination of the information needs and information sources are quite corresponding. The natural way of finding information is through books, records and files.

31. What are the two theories involved in the determination of information sources?

Ans. Redesigning of the new system from the old system and logical approach are the two theories involved in the determination of information sources.

32. Explain the logical approach.

Ans. The logical approach of determining the information sources is also known as fresh approach. The logical approach says that the detailed analysis of the existing system is not required.

33. How is a choice made between the two theories of the information source determination?

Ans. The choice of approach is done :

- It is done by the manager or designer.
- It is done on the basis of the information sources.
- It is done on the basis of the objectives of the system that is being designed.
- It is done on the preference of the old or new system.

34. What are the two types of record?

Ans. The records are of two types. One is internal and another is external record.

35. Explain internal records.

Ans. Internal records are of written type. These documents contain information about the existing system. Examples are file record, letters, reports and documents.

36. Explain external records.

Ans. External record is the record that comes from various different sources. Example is government statistics.

37. Explain the process of interview.

Ans. Interview process of managers and personnel's is a method of identifying the possible sources of information. It helps in analyzing the existing information system. The interview process helps in obtaining secure information.

38. What is the problem faced by conducting interviews? Also state its solutions.

Ans. The problem faced while conducting interviews is human made errors. The solution to overcome this error is through proper planning and increasing the confidence level.

39. Explain sampling method.

Ans. Sampling method is performed when there is a large amount of data available. In this method, only a small part of data is examined.

40. Explain the work sampling method.

Ans. The work sampling method is used to perform the following operations :

- Action of people
- Action of machine

41. What are the two techniques used in analysis and synthesis?

Ans. Analysis and synthesis makes use of two techniques to avoid the duplication. These techniques are input-output analysis and multi-dimensional flow.

42. Explain multi-dimensional flow.

Ans. Multi-dimensional flow is a technique of organizing the various information sources. This technique makes use of the flowchart. The flowchart helps in tracing the flow of information from source to destination. The multi-dimensional flowchart should be arranged in an increasing order.

43. Explain evaluation method.

Ans. The evaluation method is a method to compare the alternative designs. The method consists of the following criterion :

- The objectives of the system must be compared with the expected performance of the design of the system.
- Prepare a cost effective analysis of the system.
- Flowcharts must be examined. It helps to identify the strong and weak points of the system design.
- Study the design in more detail.
- Study the number of operations performed by the system and its potential.
- Examine the database quality.

44. What are the factors on which the documentation should be done?

Ans. Documentation should be done of :

- Flowchart
- Input
- Output
- Operations
- Flow of information

45. What is the role of manager in the documentation process?

Ans. Manager is also involved in the documentation of the system. The manager is concerned with the amount to which the system provides the required information.

46. What is the role of designer in the documentation process?

Ans. Designer is also involved in the documentation of the system. The designer is related to materials, equipment and technical processes.

47. What are the instructions that should be followed in the documentation process?

Ans. The following instructions should be followed in the documentation process :

- What data should be captured?
- When should the files be used?
- How the processing should be done?
- What output will be generated?

- How the files will be distributed?
- How the output should be distributed?

48. Explain black box Concept.

Ans. The black box concept includes the following elements *i.e.,* input, output, master files and data processing rules. In black box, the processing element remains black to the manager and user.

49. What are the rules that should be followed in the black box concept?

Ans. The rules that should be followed in the black box concept are :

- Processing of the input data against the data file and finally produce an output.
- Processing the input data for updating the file.
- Processing of input data to output data without referencing the data file.
- Producing an output from the file without processing the input.

50. Explain system flow generation.

Ans. The generation of system flow is a method of specifying the general structure of a computer based information system. It reflects the design effort.

51. Explain conceptual design report.

Ans. A conceptual design report is prepared to know the amount spend in the organizational changes. The design report must consist of the following parameters :

- Objectives
- Summary
- Nature of the system
- Reason of selecting alternative design
- Time required to design the system
- Time required to implement the system
- Resources required to design the system
- Resources required to implement the system
- Performance specifications

52. Explain the documentation of inventory management.

Ans. The documentation of the inventory management of the raw materials is concerned with the raw materials that are required for the manufacturing process. The inventory management of raw materials includes four plants. Each plant has its own computers. The three subsystems present are stock level subsystem, reorder point subsystem and economic order quantity.

Review Questions

1. Explain the conceptual design in detail.
2. Explain the defining of problems in detail.
3. Explain the system objectives in detail.
4. Explain system constraints in detail.
5. Explain determination of information needs.
6. Explain documentation and generation of reports in detail.

□□□

Section-C

Inside This Section :

➡ **Chapter-7** : Detailed System Design

Chapter 7

Detailed System Design

"How you gather, manage and use information will determine whether you win or lose."

Inside this Chapter

- Detailed System Design Overview
- Project Management of MIS Detailed System Design
- Identify Dominant and Trade-Off Criteria
- Define the Subsystems
- Sketch Detailed Operating Subsystems and Information Flow
- Determine the Degree of Automation of Each Process
- Inform and Involve the Organization Again
- Input, Output and Processing
- Early System Testing
- Software, Hardware and Tools
- Propose an Organization to Operate the System
- Document the Detailed Design
- Revisit the Manager-User

7.1. DETAILED SYSTEM DESIGN OVERVIEW

The detailed system design provides a detailed explanation of the system that has achieved the goals that were described in the conceptual system design. The detailed system design is related to the design of the operating systems. It is a step by step explanation of the procedure that is required to operate the system. This is the main **aim** of the detailed system design.

The description in the detailed system design consists of :

- Drawings in the system design
- Flowcharts
- Procedures
- Specification of the equipment's
- Specification of file
- Operating manuals that are required to run the system.
- Operating specifications

The detailed system design cannot be applied on all the systems because of the following reasons:

- Different approaches are available for designing the system.
- The design of the system is a rational process.
- The design of the system is a social process.
- The detailed explanation may not give an idea about the general nature of design.

- The designer must interfere frequently with the detailed system design with the help of descriptive essays.

7.2. PROJECT MANAGEMENT OF MIS DETAILED SYSTEM DESIGN

Planning and organizing : The first step that is performed in the detailed system design is planning and organizing.

The project management is done for both small and large projects.

- **Small projects :** All the phases must be planned before performing the conceptual system design.
- **Large projects :** All the phases must be planned after the conceptual system design.

The project planning that is done in advance goes wasted, if it is shown by the conceptual system design that the system design is not appropriate.

Amount of effort : The amount of effort that is required in the project management of the detailed system design depends upon :

- Size of the MIS project
- Cost

7.2.1. Project Planning

The project planning of the detailed system design consists of :

(*a*) **Establish the project objectives :** The establishment of project objectives consists of :

- Review of the objectives
- Subdivision of the objectives
- Modification of the objectives

These are the objectives that were formed in the conceptual system design.

(*b*) **Define the project tasks :** Here, the tasks that are to be performed are identified. These tasks are also documented.

(*c*) **Plan :** Planning of sequential and concurrent tasks is performed. A network diagram showing the events and activities is required.

(*d*) **Schedule :** As per the requirements of the management, scheduling should be performed. End date and network constraints should be established.

(*e*) **PERT diagram :** The work and schedule are tied together with the help of a PERT diagram.

(*f*) Estimate the labor, equipment and cost required in the project.

(*g*) **Budget :** A budget should be made for the project. Funds should be allocated to each task.

(*h*) Plan the staffing organization.

7.2.2. Project Control

In the controlling of project, following factors are taken into consideration:

(*a*) During the progress of the project, determine whether the project objectives are fulfilled or not.

(*b*) **The expenditure of the project must be evaluated in terms of :**

- Time
- Work done

The budget must be revised if required.

(*c*) Evaluate the work force utilization and work performed by an individual.

(*d*) Evaluate time, cost and work requirements in terms of :

- Schedule
- Budget

7.3. IDENTIFY DOMINANT AND TRADE-OFF CRITERIA

The **dominant criterion** is a criterion that makes an activity important over other activities. The dominant trade off criteria must be identified before undergoing the design steps. For example : Company selling life preserving drugs, company dealing with electric power etc.

The **trade-off criterion** is the criterion in which performance of one activity is reduced to increase the performance of another activity. For example : Criteria of low manufacturing cost are balanced with the public image of the organization that is achieved by reducing the pollution level.

> **DO YOU KNOW?**
>
> It's not the faith in technology; it's the faith in people.
>
> **–Steve Jobs**

Identification reason of dominant and trade-off criteria : There are two reasons for identifying the dominant and trade-off criteria. These are :

- The decision centers must be identified when the detailed design is developed. It helps to achieve the dominant and trade off criteria.
- A proper management information system must be developed to provide relevant information. It helps in taking decisions regarding the trade-off.

7.4. DEFINE THE SUBSYSTEMS

The defining of subsystems is done with the help of two main blocks of information. These are :

1. **Conceptual Design :** Conceptual design requires assumptions regarding the subsystems. The conceptual design contains :
   - Review of subsystems
   - Redefining of subsystems if required
   - Detailed activity of each important activity block
2. **Dominant and trade-off performance criterion :** An activity design form is used to develop description about each activity. Trade-off criteria must be used when dominant criteria is not present. The **trade-off criteria** stresses on different functional activities thus helping in defining the subsystems in a more proper and relevant manner.

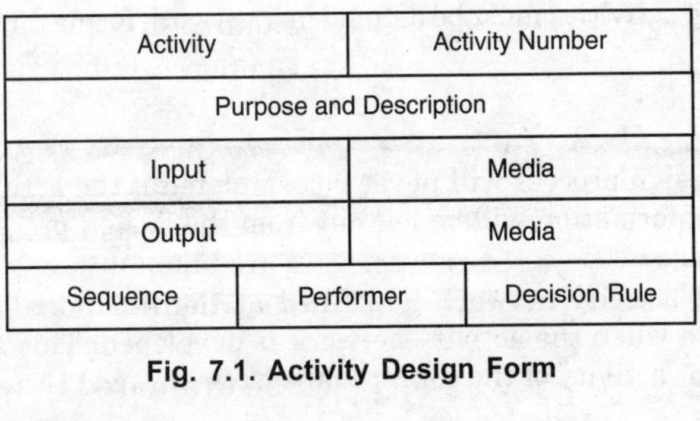

Fig. 7.1. Activity Design Form

7.4.1. Quality Control Activity

The quality control activity consists of :

- **Activity Processor** : The activity processor consists of the equipment's and the number of workers assigned to perform a particular task.
- **Information Output :** It is the resources available after assigning to a particular activity.
- **Operating Output** : Batches are created. The operating output is the reduced size batches.
- **Information Results** : The information result provides information about the average quality of batches and the number of batches that are inspected in a unit of time.

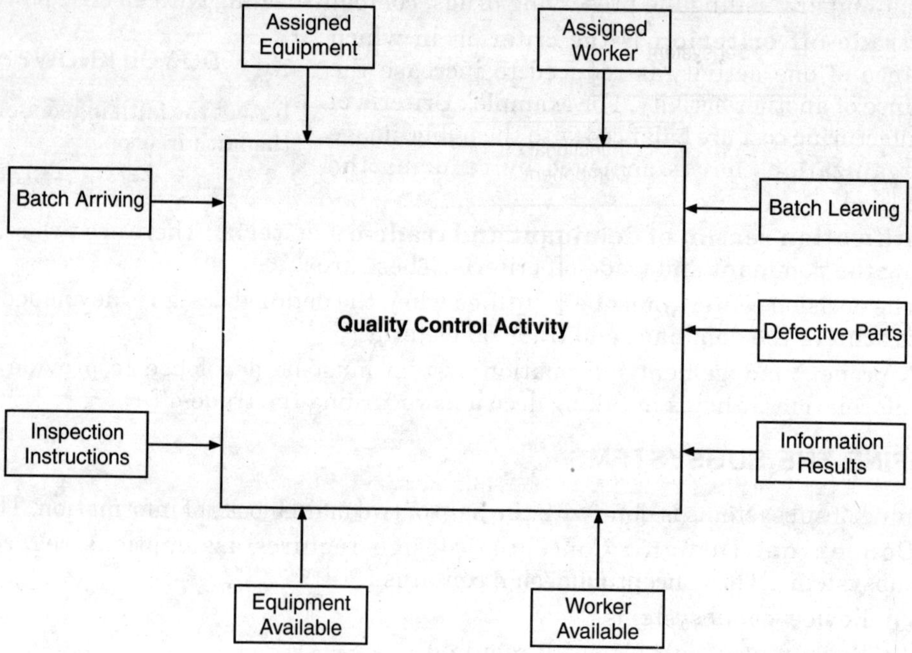

Fig. 7.2. Quality Control Activity

7.4.2. Activity

The activities can be explained in two terms : Inter-relating activities and hierarchy of activities.

The **inter-relating** activities must be defined in a network format. The degree of breakdown of the activities shows :

- Size
- Complexity

of the network. The design process will never be completed if the activities are broken down too finely. Important information will be left out from the design process if the activities are broken down too broadly.

The **hierarchy** of activity network is defined at the functional component level. The subsystems are defined when the activity network is developed. This activity network takes into account each major activity of the conceptual system design.

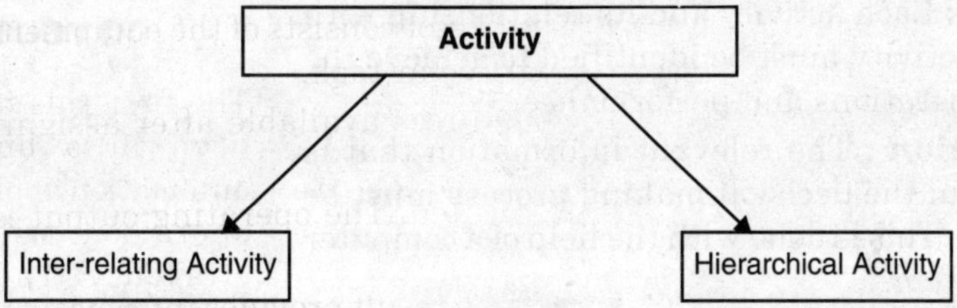

Fig. 7.3. Activity

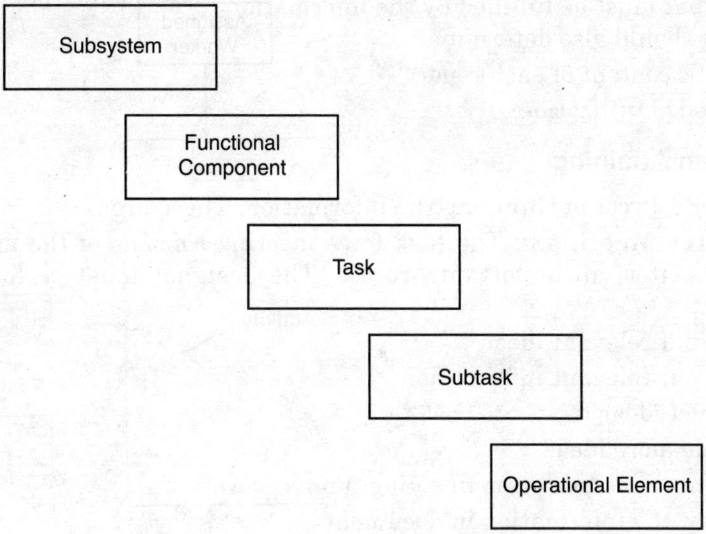

Fig. 7.4. Activity Hierarchy

The **subsystem** consists of :

- Activities matching to major activity block of the conceptual system design.
- **Detailed activity blocks :** These can be transferred from one group to another so that a network of subsystem can be created.

But now the question arises to how to identify the new subsystem. So, the new revised sub-system can be identified by drawing a loop around the group of activities that are to be included in the subsystem.

Grouping of Activities : The grouping of activities is done on the basis of the following factors :

- Common functions
- Common techniques
- Common inputs
- Common outputs
- Logical flow relationship

7.4.3. Information Components

The information must consist of the following factors. These are :

(*a*) Available Resources : The system must be developed in terms of the resources that are available in the company.

(*b*) **Activity :** Each activity and its relationship with another activity must be identified to achieve the system operations and performance.

(*c*) **Information :** The relevant information that is required in the decision making process must be identified. This is done with the help of a computer based system.

(*d*) **Output :** A specific output is required by all the systems. This should include a detailed list of purpose that must be fulfilled by the information output. It should also determine:

- Specific content of each report
- Method of utilization

DO YOU KNOW?

The first rule of any technology used in a business is that automation applied to an efficient operation will magnify the efficiency. The second is that automation applied to an inefficient operation will magnify the inefficiency.

–Bill Gates

7.4.4. Information Attaining

There are four sources of obtaining the information. These are :

(*a*) **Task Force Meetings :** The task force meeting consists of the managers and key specialists. It is an important process. The designer must perform the following functions :

- Draw out relevant ideas
- Draw out relevant information
- Abstract ideas
- Include more ideas
- Present the information in a diagrammatic way
- Present the information in documents
- Modify the information
- Evaluate the information

The purpose of the task force meetings is :

- Reduce the information gap
- Increase the operating needs
- Decrease debatable points

(*b*) **Interviews :** The interview process is conducted by the manager and the designer to identify objectives and develop a proper plan. This is performed at both the top and intermediate level. It is in addition to the task force meetings. Employees of the organization and specialists should also take part in the interview process.

The designer should search for information on :

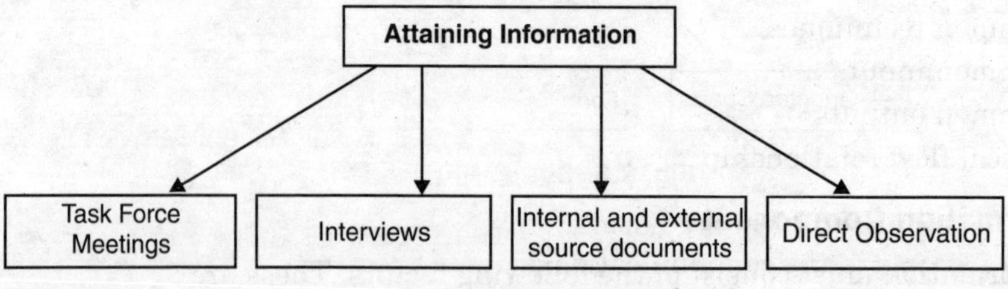

Fig. 7.5. Attaining Information

- Objectives of the organization

- Format of the information that is presented
- Style of manager on the decision making process
- Resources required for operation of the system
- Resources required for implementation of the system
- Degree of manager's involvement in the decision making process

(c) Internal and External Source Documents : The source documents can either be internal or external.

The **internal** source document provides the system designer with the point of withdrawal. The internal source documents may be large depending upon the size of the company. The traditional reports and documents must be retained.

For Example : Procedure manuals, record of customer complaints.

The **external** document provides :

- Economic information
- Marketing information
- Industrial information
- Financial information

(d) Direct Observation : Direct observation also plays an important role in obtaining the information. Designer should also perform on the spot surveys to gather more information and record the relevant data simultaneously. It helps in knowing the physical and environmental factors that are to be accepted in the new system. The designer should organize the data at the end of phase.

7.5. SKETCH DETAILED OPERATING SUBSYSTEMS AND INFORMATION FLOW

The operating subsystem and operating system must be sketched in a detailed manner. The detailed system design must be carried in the following order :

- Subsystem level
- Functional level
- Task level

There are three types of system flowcharts. These are :

(a) Task oriented charts : Task oriented charts are basically the block diagrams that show the relationship among various tasks. A step by step analyzing is done on the steps that are required to complete an activity.

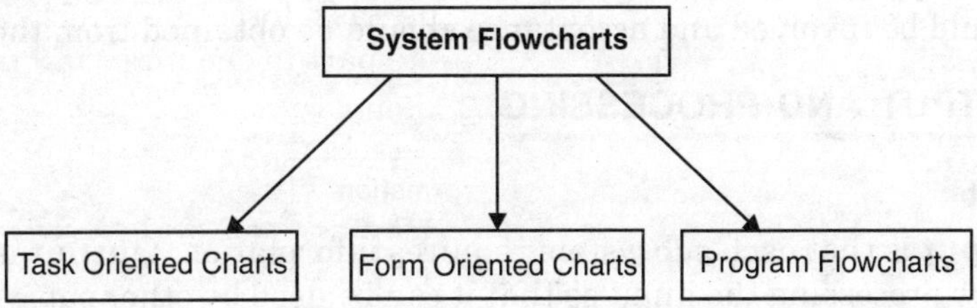

Fig. 7.6. System Flowcharts

(b) Forms oriented charts : Form oriented charts are the ones that identify the forms that are used in the communication process. It traces the flow of all the forms through the organization.

(c) Program flowcharts : Program flowcharts are created by the people who provide instructions to the computer. These flowcharts are the fundamental tool of the programming design. It is a logical sequence of steps that must be done by the computer.

7.6. DETERMINE THE DEGREE OF AUTOMATION OF EACH PROCESS

The management can automate the processes to different degrees according to its wish and needs. The degree of automation can be classified as :

(*a*) **No automation :** Manual processes are used by the people to perform the functions.

(*b*) **Work automated, manual control :** In this, the work is done by the electromechanical devices but the control is performed manually.

(*c*) **Work automated, manual control, automated feedback :** The work is done by the electromechanical devices but the control is performed manually. There is a feedback loop that is controlled by the devices and gives information to the operator.

(*d*) **Work automated, automated control, automated feedback :** In this no human control is required.

Low level automation is preferred in the following cases :

- Problems are not well structured
- Decision criteria is not well defined
- Gaps in data that is entered into the system
- Cost of human labor is low
- Negligible storage
- Few processing steps
- Processing steps are simple
- Data entered into the system is inconsistent

7.7. INFORM AND INVOLVE THE ORGANIZATION AGAIN

The organization must be involved and informed. The management can be classified as upper management and lower management.

Upper level management : The upper management should be provided with a brief overview of the design and status of the design effort.

Lower level management : The lower level management should be provided with :

- System flowchart
- Subsystem flowchart

The management should comment on the accuracy of the system. It should also guide on the amount of automation required. It helps us to know about the nature of the managers i.e. whether they are supportive or resistive. The informing and involving process also requires that the user should be involved and acceptance should be obtained from the user side.

7.8. INPUT, OUTPUT AND PROCESSING

Input and Output

FORMS : The **input** is that each subsystem requires information. **Output** is the information that is produced by processing the input so that it can be used by other subsystems. The input and output operations are normally done with the help of **forms.** Form is a communication link that is used in the operating procedure.

The major tasks in performing this operation are :

- Input should be specified properly
- Output should be specified properly
- Design the forms that are relevant

- Design the forms that are easy to use

Identification Task

- Functions of the form
- No of copies required
- Where is the form to be used?
- Who will use the form?
- How many units of form are required per year?

Form Designing Factors : The following factors should be considered while designing the forms :

- Forms should be attractive
- Forms should be easy to read and use
- Forms should contain an identification number
- Forms should contain the date of issue
- Forms should be designed in a logical manner either from left to right or top to bottom
- Forms should contain sufficient data to enter the data

Input and Output

DATABASE : Database is a data that can be stored and retrieved. It helps in making decisions. The decisions can either be managerial decisions or programmed decisions.

Approach : The approach that must be followed in the development of database is :

- Identification of all the points in the flowchart that require input data.
- Prepare data for each data element providing the :
  - Source of data
  - Length of data
  - Format of data
  - End use of data
- Group all the data with activity
- Check for errors
- Check if any duplicate data is present
- Eliminate the duplicate data so that an integrated database is developed

Processing : In the processing following factors are there :

- Vast amount of information has been gathered
- Input is present in a detailed manner
- Output is present in a detailed manner
- Corrections should be made to the subsystem
- The task that is specified in the subsystem should be broken down either in a mathematical manner or narrative format.

7.9. EARLY SYSTEM TESTING

The early system testing consists of three parts, *i.e.*, modeling, simulation and test planning.

(*a*) **Modeling :** Modeling method is used to define the systems in a more proper manner. The modeling method is used for the following functions :

- Determine the quantitative range of input
- Determine the quantitative range of output

- Establish a relationship for performing the transfer functions
- Timely response for the operations in the system

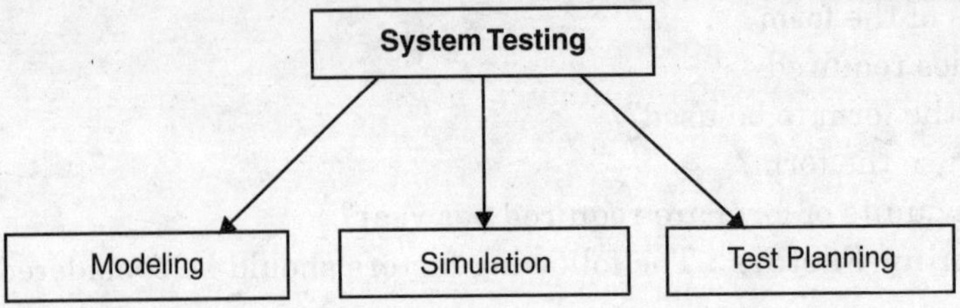

Fig. 7.7. System Testing

The modeling process consists of **decision tables**. The decision table consists of :

– Quantitative basis
– Qualitative basis

for the decision making process. The decision table is required for both design and implementation of the system.

Format of decision tables : The structure of the decision tables is :

If these conditions exist
Then perform these actions

(b) Simulation : Testing by simulation is carried out on the :

- System
- Subsystem
- Functional components

Small systems : The testing functions in the smaller organizations are carried out by the conversion to online operations.

Large systems : The testing function in large organizations is quite complex and costly.

Simulation Procedure : The procedure for simulation is carried out in the following steps :

1. Select the value of exogenous data like interest rate, cost of raw materials randomly.
2. Input that is non-quantitative in nature is handled by listing all the alternatives and selecting a random value.
3. Trace the effect of exogenous input data throughout the system.
4. Examine the output of various subsystems.
5. Repeat steps 1 to 3 different times.

Activity Process : The activity process can either be automatic or manual.

1. **Manual Activity Process :** In the manual automatic process, errors may occur.
2. **Automatic Activity Process :** The automatic activity process is direct. For example : Computerized systems, decision tables etc.

(c) Test Planning : Individuals show the detailed plans for verifying the final version of the management information system. In the test planning, the following mentioned questions should be answered in a detailed manner :

- What are the user interfaces?
- Which subsystem will be available on different dates?
- Who are the users?
- Who are the managers?

7.10. SOFTWARE, HARDWARE AND TOOLS

Software, hardware and tools consist of steps that must be followed. It is the responsibility of the system designer to check these steps in a proper manner.

Steps for software : The steps for software are :

- Develop standards for programming
- Develop programs for programming
- Develop the logic
- Develop programming flowcharts
- Study the conceptual system design specifications
- Work with the system designer in a team to develop detailed system design.

Steps for hardware : The steps for hardware are :

- Enough computing power must be purchased to perform the whole job
- Enough external storage must be purchased to store the required database
- Peripheral devices must be purchased to support the MIS in solving the problems
- Computer supplies must be purchased. For example : paper, tapes etc
- Best documentation must be purchased
- Maintenance package must be purchased

7.11. PROPOSE AN ORGANIZATION TO OPERATE THE SYSTEM

An organization must be proposed to operate the system. The management information system group should work with the top managers to suggest changes that will fulfill the requirements of the new system. It is the responsibility of the subsystem managers that the subsystem should work in the best possible way. They should also try to increase the performance of the system.

7.12. DOCUMENT THE DETAILED DESIGN

Documents' specifying the system, operation and design is the end product of the detailed system design. The documentation of the detailed system design must consist of :

- Summary flowchart
- Detailed flowchart
- Specification of database
- Hardware requirements
- Software requirements
- Performance specifications
- Cost of operations

The documentation should be in the following format :

- Standardized format
- Input and output activity must be linked
- Good reporting style should be used

7.13. REVISIT THE MANAGER-USER

The following operations should be performed :

- Report the status to the upper management
- Request the lower management for support

- Request the lower management for providing feedback
- Educate the non-management user about the system
- Sell the system to the non-management user

Summary

- The detailed system design is a step by step explanation of the procedure that is required to operate the system.
- The first step that is performed in the detailed system design is planning and organizing.
- The project management is done for both small and large projects.
- The amount of effort that is required in the project management of the detailed system design depends upon the size of the MIS project and its cost.
- Planning of sequential and concurrent tasks is performed.
- The dominant criterion is a criterion that makes an activity important over other activities.
- The trade-off criterion is the criterion in which performance of one activity is reduced to increase the performance of another activity.
- The defining of subsystems is done with the help of two main blocks of information. These are conceptual design and dominant and trade off criteria.
- The trade-off criteria stresses on different functional activities thus helping in defining the subsystems in a more proper and relevant manner.
- The quality control activity consists of an activity processor, information output, operating outputs and information results.
- The activity processor consists of the equipment's and the number of workers assigned to perform a particular task.
- The information result in a quality control activity provides information about the average quality of batches and the number of batches that are inspected in a unit of time.
- The inter-relating activities must be defined in a network format. The degree of breakdown of the activities shows the size and complexity of the network.
- The hierarchy of activity network is defined at the functional component level.
- The system must be developed in terms of the resources that are available in the company.
- The task force meeting consists of the managers and key specialists.
- The interview process is conducted by the manager and designer to identify objectives and develop a proper plan.
- The source documents can either be internal or external.
- The internal source document provides the system designer with the point of withdrawal.
- Direct observation plays an important role in obtaining the information. Designer should also perform on the spot surveys to gather more information and record the relevant data simultaneously.
- The operating subsystem and operating system must be sketched in a detailed manner.
- Task oriented charts are basically the block diagrams that show the relationship among various tasks. A step by step analysing is done on the steps that are required to complete an activity.
- Form oriented charts are the ones that identify the forms that are used in the communication process.
- Program flowcharts are created by the people who provide instructions to the computer.

- The management can automate the processes to different degrees according to its wish and needs.
- The upper management should be provided with a brief overview of the design and status of the design effort.
- The lower level management should be provided with system flowchart and subsystem flowchart.
- Form is a communication link that is used in the operating procedure.
- Database is a data that can be stored and retrieved. It helps in making decisions. The decisions can either be managerial decisions or programmed decisions.
- Modeling method is used to define the systems in a more proper manner.
- The modeling process consists of decision tables. The decision table consists of quantitative and qualitative basis for the decision making process.
- The testing functions in the smaller organizations are carried out by the conversion to online operations.
- The testing function in large organizations is quite complex and costly.
- The management information system group should work with the top managers to suggest changes that will fulfill the requirements of the new system.
- Documents' specifying the system, operation and design are the end product of the detailed system design.

Short Question Answers

1. Give an overview of detailed system design.

Ans. The detailed system design provides a detailed explanation of the system that has achieved the goals that were described in the conceptual system design. The detailed system design is related to the design of the operating systems. It is a step by step explanation of the procedure that is required to operate the system.

2. What does the detailed system design consists of?

Ans. The description in the detailed system design consists of :

- Drawings in the system design
- Flowcharts
- Procedures
- Specification of the equipment's
- Specification of file
- Operating manuals that are required to run the system.
- Operating specifications

3. Why cannot the detailed system design applied to all systems?

Ans. The detailed system design cannot be applied on all the systems because of the following reasons :

- Number of different approaches is available for designing the system.
- The design of the system is a rational process.
- The design of the system is a social process.
- The detailed explanation may not give an idea about the general nature of design.
- The designer must interfere frequently with the detailed system design with the help of descriptive essays.

4. What is the first step in the detailed system design?

Ans. The first step that is performed in the detailed system design is planning and organizing. The project management is done for both the small and large projects.

5. **Explain the project management for projects?**

Ans. The project management is done for both small and large projects.

- **Small projects :** All the phases must be planned before performing the conceptual system design.
- **Large projects :** All the phases must be planned after the conceptual system design.

6. **What are the components of project planning?**

Ans. Establish the project objectives, define the project tasks, plan, schedule, PERT diagram and budget are the components of project planning.

7. **What is a dominant criterion?**

Ans. The dominant criterion is a criterion that makes an activity important over other activities. The dominant trade off criteria must be identified before undergoing the design steps. For example : Company selling life preserving drugs, company dealing with electric power etc.

8. **What is trade-off criterion?**

Ans. The trade-off criterion is the criterion in which performance of one activity is reduced to increase the performance of another activity. For example : Criteria of low manufacturing cost are balanced with the public image of an organization that is achieved by reducing the pollution level.

9. **What are the identification reason of dominant and trade-off criteria?**

Ans. There are two reasons for identifying the dominant and trade-off criteria. These are :

- The decision centers must be identified when the detailed design is developed. It helps to achieve the dominant and trade off criteria.
- A proper management information system must be developed to provide relevant information. It helps in taking decisions regarding the trade-off.

10. **Explain the defining of subsystems?**

Ans. The defining of subsystems is done with the help of two main blocks of information. These are :

1. **Conceptual Design :** Conceptual design requires assumptions regarding the subsystems.
2. **Dominant and trade-off performance criterion :** An activity design form is used to develop description about each activity. Trade-off criteria must be used when dominant criteria is not present. The trade-off criteria stresses on different functional activities thus aiding in defining the subsystems in a more proper and relevant manner.

11. **What are the components of quality control activity?**

Ans. Activity processor, information output, operating output and information results are the components of quality control activity.

12. **What do you mean by activity processor?**

Ans. The activity processor consists of the equipment's and the number of workers assigned to perform a particular task.

13. **What do you mean by information results?**

Ans. The information result provides information about the average quality of batches and the number of batches that are inspected in a unit of time.

14. **What are the two types of activity?**

Ans. Inter-relating activity and hierarchy activity are the two types of activity.

15. **What do you mean by inter-relating activity?**

Ans. The inter-relating activities must be defined in a network format. The degree of breakdown of the activities shows size and complexity of the network.

16. **What do you mean by hierarchy of activity?**

Ans. The hierarchy of activity network is defined at the functional component level. The subsystems are defined when the activity network is developed.

17. What does the subsystem consists of?

Ans. The subsystem consists of :

- Activities matching to major activity block of conceptual system design.
- Detailed activity blocks : These can be transferred from one group to another so that a network of subsystem can be created.

18. How is the new subsystem identified?

Ans. The new revised sub-system can be identified by drawing a loop around the group of activities that are to be included in the subsystem.

19. How is the grouping of activities done?

Ans. The grouping of activities is done on the basis of the following factors :

- Common functions
- Common techniques
- Common inputs
- Common outputs
- Logical flow relationship

20. What are the four sources of obtaining information?

Ans. Task force meetings, interviews, direct observation, internal and external documents are the four sources of obtaining information.

21. Explain task force meetings.

Ans. The task force meeting consists of the managers and key specialists. It is an important process.

22. What are the functions that must be performed by the designer in the task force meeting?

Ans. The designer must perform the following functions :

- Draw out relevant ideas
- Draw out relevant information
- Abstract ideas
- Include more ideas
- Present the information in a diagrammatic way
- Present the information in documents
- Modify the information
- Evaluate the information

23. What is the purpose of task force meeting?

Ans. The purpose of the task force meetings is :

- Reduce the information gap
- Increase the operating needs
- Decrease debatable points

24. Explain interview process.

Ans. The interview process is conducted by the manager and the designer to identify objectives and develop a proper plan. This is performed at both the top and intermediate level. It is in addition to the task force meetings. Employees of the organization and specialists should also take part in the interview process.

25. What information should the designer search in the interview process?

Ans. The designer should search for information on :

- Objectives of the organization
- Format of the information that is presented
- Style of manager on the decision making process
- Resources required for operation of the system

- Resources required for implementation of the system
- Degree of manager's involvement in the decision making process

26. What do you mean by internal documents?

Ans. The internal source document provides the systcm designer with the point of withdrawal. The internal source documents may be large depending upon the size of the company. The traditional reports and documents must be retained. For Example : Procedure manuals, record of customer complaints.

27. What information does the external document provide?

Ans. The external document provides :

- Economic information
- Marketing information
- Industrial information
- Financial information

28. What do you mean by direct observation?

Ans. Direct observation also plays an important role in obtaining the information. Designer should also perform on the spot surveys to gather more information and record the relevant data simultaneously. It helps in knowing the physical and environmental factors that are to be accepted in the new system. The designer should organize the data at the end of phase.

29. What are the three types of system flowcharts?

Ans. Task oriented charts, forms oriented charts and program flowcharts are the three types of system flowcharts.

30. Explain task oriented charts.

Ans. Task oriented charts are basically the block diagrams that show the relationship among various tasks. A step by step analyzing is done on the steps that are required to complete an activity.

31. Explain forms oriented charts.

Ans. Form oriented charts are the ones that identify the forms that are used in the communication process. It traces the flow of all the forms through the organization.

32. Explain program flowcharts.

Ans. Program flowcharts are created by the people who provide instructions to the computer. These flowcharts are the fundamental tool of the programming design. It is a logical sequence of steps that must be done by the computer.

33. Classify the degree of automation.

Ans. The degree of automation can be classified as :

(*a*) **No automation :** Manual processes are used by the people to perform the functions.

(*b*) **Work automated, manual control :** In this, the work is done by the electromechanical devices but the control is performed manually.

(*c*) **Work automated, manual control, automated feedback :** The work is done by the electromechanical devices but the control is performed manually. There is a feedback loop that is controlled by the devices and gives information to the operator.

(*d*) **Work automated, automated control, automated feedback :** In this no human control is required.

34. What are the cases in which low level automation is preferred?

Ans. Low level automation is preferred in the following cases :

- Problems are not well structured
- Decision criteria is not well defined
- Gaps in data that is entered into the system
- Cost of human labor is low

- Negligible storage
- Few processing steps
- Processing steps are simple
- Data entered into the system is inconsistent

35. Classify management.

Ans. The management can be classified as upper management and lower management. The upper management should be provided with a brief overview of the design and status of the design effort. The lower level management should be provided with system flowchart and subsystem flowchart.

36. What are the forms that must be considered while designing the forms?

Ans. The following factors should be considered while designing the forms :

- Forms should be attractive
- Forms should be easy to read and use
- Forms should contain an identification number
- Forms should contain the date of issue
- Forms should be designed in a logical manner either from left to right or top to bottom
- Forms should contain sufficient data to enter the data

37. What is the approach followed in the development of database?

Ans. The approach that must be followed in the development of database is :

- Identification of all the points in the flowchart that require input data.
- Prepare data for each data element providing the :
  - Source of data
  - Length of data
  - Format of data
  - End use of data
- Group all the data with activity
- Check for errors
- Check if any duplicate data is present
- Eliminate the duplicate data so that an integrated database is developed

38. What is a database?

Ans. Database is a data that can be stored and retrieved. It helps in making decisions. The decisions can either be managerial decisions or programmed decisions.

39. What are the three parts of early system testing?

Ans. The early system testing consists of three parts i.e. modeling, simulation and test planning.

40. Explain modeling.

Ans. Modeling method is used to define the systems in a more proper manner. The modeling process consists of decision tables. The decision table consists of quantitative basis and qualitative basis for the decision making process. The decision table is required for both design and implementation of the system.

41. What are the functions for which the modeling method is used?

Ans. The modeling method is used for the following functions :

- Determine the quantitative range of input
- Determine the quantitative range of output
- Establish a relationship for performing the transfer functions
- Timely response for the operations in the system

42. What is the structure of decision table?

Ans. The structure of the decision tables is :

if these conditions exist
then perform these actions

43. What do you mean by simulation?

Ans. Testing by simulation is carried out on the system, subsystem and functional components.

44. What is the simulation procedure?

Ans. The procedure for simulation is carried out in the following steps :

1. Select value of exogenous data like interest rate, cost of raw materials randomly.
2. Input that is non-quantitative in nature is handled by listing all the alternatives and selecting a random value.
3. Trace the effect of exogenous data input throughout the system.
4. Examine the output of various subsystems.
5. Repeat steps 1 to 3 different times.

45. Explain activity process.

Ans. The activity process can either be automatic or manual.

1. **Manual Activity Process** : In the manual automatic process, errors may occur.
2. **Automatic Activity Process :** The automatic activity process is direct. For example : Computerized systems, decision tables etc.

46. Explain test planning.

Ans. Individuals show the detailed plans for verifying the final version of the management information system.

47. What are the steps for software?

Ans. The steps for software are :

- Develop standards for programming
- Develop programs for programming
- Develop the logic
- Develop programming flowcharts
- Study the conceptual system design specifications
- Work with the system designer in a team to develop detailed design

48. What are the steps for hardware?

Ans. The steps for hardware are :

- Enough computing power must be purchased to perform the whole job
- Enough external storage must be purchased to store the required database
- Peripheral devices must be purchased to support the MIS in solving the problems
- Computer supplies must be purchase. For example : paper, tapes etc
- Best documentation must be purchased
- Maintenance package must be purchased

49. What must be documented in the detailed system design?

Ans. The documentation of the detailed system design must consist of :

- Summary flowchart
- Detailed flowchart
- Specification of database
- Hardware requirements
- Software requirements
- Performance specifications
- Cost of operation

50. What format should be followed in the documentation process?

Ans. The documentation should be in the following format :

- Standardized format
- Input and output activity must be linked
- Good reporting style should be used

Review Questions

1. Give an overview of the detailed system design.
2. Explain inform and involve the organization.
3. Explain the aim of detailed system design.
4. Explain project management of MIS.
5. Explain the identification of dominant and trade-off criterion.
6. Explain define the subsystems.
7. Sketch the detailed system operating and information flow.
8. Determine the degree of automation of each operation.
9. Explain informing and involving of organization.
10. Explain input, output and processing.
11. Explain early system testing.
12. Explain software, hardware and tools.
13. Propose an organization to operate the system. Explain.
14. Explain the documentation of detailed system design.
15. Explain the revisiting the manager and user again.

□□□

Section-D

Inside This Section :

Chapter 8

Implementation of Management Information System

"We may ask for information, but we are usually only interested in what confirms are opinions."

Inside this Chapter

A system design is of no use until it is not put into operation. Once the system's designing process is over, its blue print is available in the form of design description tools. Now the next phase in the development cycle is to ensure that the design process is over. But sometimes both the design and the implementation phases may overlap each other.

Here, the role of implementation phase steps in. In this chapter you will learn about the issues related to the implementation, evaluation and maintenance of a system.

8.1. IMPLEMENTATION

Definition : Implementation is an important managerial activity related to the adoption and management of an innovation. It is an activity which starts when the system is developed and is ready to be installed in the organization.

The design of the Management Information system seems to the management as an expensive project. The implementation of MIS is done when the output of the MIS are continuously used by the decision makers. So basically implementation is a process of introducing a newly developed Management Information System into an organization so that it can be used by the end user, owners etc.

✍ **Note :** In the absence of a proper implementation system, the most sophisticated system will also be a failure.

Steps before the implementation process occur : The appropriate steps that must be taken before the implementation process is started are :

- Designer must make an appropriate environment so that he gets full support from all the levels of management in an organization.
- The user requires adequate training so that he can learn to implement the system efficiently and effectively.

The implementation outcome can be subjected to various factors like :

- Role of users in the implementation process.
- Amount of management support for the implementation process.
- Complexity of risk involved in the implementation activity.
- Level of risk involved in the implementation activity.
- Quality of management in the implementation process.

System analyst : System analyst is the catalyst that is responsible for the entire change of process and makes sure that the changes are acceptable to all the persons that are working in an organization. Thus, the system analyst plays the role of a change agent.

8.2. APPROACHES TO ACHIEVE IMPLEMENTATION

The implementation of MIS can be achieved by using any one of the four methods which are discussed below. These are :

- Direct Approach
- Parallel Approach
- Modular Approach
- Phase In Approach

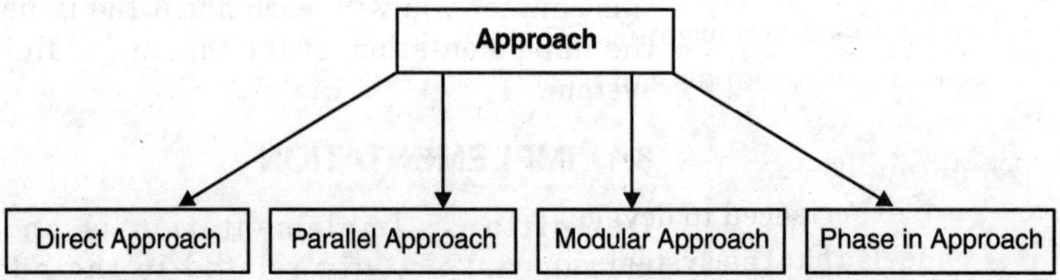

Fig. 8.1. Approaches to achieve Implementation

8.2.1. Direct Approach

In the direct approach, the new system is installed directly with the immediate discontinuance of the old existing system. It is also known as immediate cut over or change over. Direct approach is useful when the below mentioned factors are taken into consideration:

- The new system does not replace the existing system.
- The old system is now of zero value.
- The new employing system is compact.
- The new system is simple.
- The design of the new system is inexpensive.
- The new system involves fewer risks and has more advantages than the old one.

✍ **Note:** The direct approach is also referred to as **"COLD TRUNKY"** approach.

8.2.2. Parallel Approach

In the parallel approach, the selected new system is installed and operated with the current system in parallel.

Features :

- Expensive approach: It is expensive because of the duplicating services and the personnel required to maintain both the old and new system.

In parallel approach, a target date must be fixed to know when the operations of the old system end and the new system starts operating on its own.

8.2.3. Modular Approach

The modular approach is also known as **"PILOT APPROACH"**. In this approach, the new system is introduced at one location only for the trial purpose. If its performance is according to the needs of an organization, it is introduced in the entire organization otherwise not. Thus, implementation of the MIS system in the organization is done on peace-meal basis.

Advantages :

- The major problem can be easily identified and corrected before further implementation of the system is done.
- The operating personnel can be trained before the system is installed in the location.
- The risk of systems failure is limited.

8.2.4. Phase in Implementation

The phase in implementation is quite similar to the modular approach. The only difference between the two is that; phase in implementation deals with the segmentation of the system whereas the modular approach deals with the segmentation of an organization.

Advantages :

- Rate of changes in a given organization can be totally minimized.
- The data processing resources can be acquired gradually over a period of time.

Disadvantages :

- Limited applicability
- More costs are experienced to develop an interface with the old system
- Organization feels that the system is never completed

8.3. METHODS OF IMPLEMENTING MIS

After the designing phase is over, there are four methods of implementing a Management Information System. These are :

1. Install the new system in the new organization.
2. **Cut off the old system and install a new system :**
   - During the cut off of the old system and installation of a new system, a time gap is produced during which no system is in operation.
   - This method is practical only for small companies where installation requires only one or two days.
   - The exception of this approach can be the installation of systems in larger companies during vacation or period of inactivity.

3. Cut over by segments :

- This method is also referred as "phasing in" the new system.
- Small parts or subsystems are substituted for the old system.
- In the case of upgrading old systems, this may be a very desirable and useful method.

4. Operate in parallel and cut over :

- The new system is installed and operated in parallel with the current system. The current system is cut out when the new system is checked out properly.
- Expensive method due to personnel and related costs.
- The advantage of using this method is that the system is well debugged.

Example : Payroll system, Customer Billing.

8.4. IMPLEMENTATION STEPS OF A MANAGEMENT INFORMATION SYSTEM

The implementation of the Management Information System is the result of the designing process. The documentation of the system is necessary during the installation of the system so that there is an up to date reference. The final documentation should be :

- Complete
- Formal
- Accurate version of the Management Information System

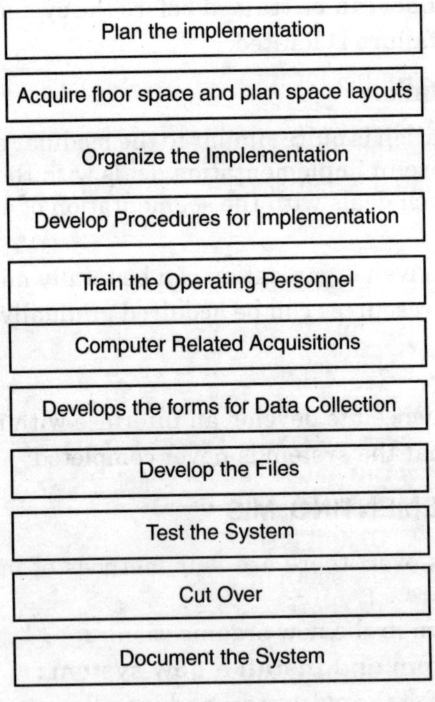

Fig. 8.2. Implementation steps of MIS

8.4.1. Plan the Implementation

Planning the implementation of MIS in an organization is the first step in the implementation phase. In planning the implementation of the MIS, there are three main phases which occurs in a series. These three phases are listed below :

- Initial Installation
- Test of the whole system
- Evaluation, maintenance and control of the system

The planning process is a function of line management. The systems analyst must prepare plans to assist the managers but the managers are involved in last, only if required. Also, the manager requires system analyst to make detailed plans. The manager prefers to make decisions by using the most recent information. This recent information includes :

- Management Information System Specification
- Proposed plans of the System Analyst
- Current operating situation

The planning phase of the implementation process must include the following :

- Conclusion of the design process
- Documentation of the system

Many implementation activities should be started in parallel to reduce the implementation time.

For Example : Training of personnel and preparation of software can be done in parallel with each other and with other implementation activities.

(*a*) **Identify the Implementation Tasks :** The major implementation tasks consists of :

- Planning the implementation activities
- Acquiring and laying out facilities and offices
- Organizing the personnel for implementation
- Developing procedures for installation and testing
- Developing the training program for operating personnel
- Completing the system's software
- Acquiring required hardware
- Generating files
- Designing forms
- Testing the entire system
- Complete cutover to the new system
- Documenting the system
- Evaluating the MIS
- Providing system maintenance (debugging and improving)

(*b*) **Establish Relationships among Tasks :** The establishment of relationship among tasks is done for both small and large projects.

- **Small Projects :** For the small projects, the order of performance may be described in a text format. A Gantt chart or network diagram makes visualization of the plan and schedule much clearer.
- **Large Projects :** For large projects, many parallel and sequential activities are interrelated so that a network diagram can also be used.

(*c*) **Establishing of MIS :** Management Information System should be established because of the following reasons :

- Monitoring the progress of Implementation
- Proper control of activities
- Efficient Information system should be developed.

(*d*) **Establish a Schedule :** Schedule is prepared by the system designers. The role of the system designer is to :

- Estimate the time between the events in a program network.
- The critical path that is the longest time through the network can also be calculated.
- End date is established by knowing the starting date.

(*e*) **Prepare a Cost Schedule tied to tasks and time :** The cost of completing each task is established as part of the plan. The rate of expenditure should also be calculated.

(*f*) **Establishing a Reporting and Control system :** Establishing of a Reporting and Control system is performed by the key people involved.

- Reporting of the work in progress should be obtained in weekly meetings.
- Control of the work in progress may be obtained by weekly meetings.
- Financial personnel must make sure that report formats allow them to show relationship between cost and the technical progress.

The report establishes two relationships :

- Cost and time relationship
- Cost and technical progress relationship

The objective of control system is to minimize the confusion, delay and costs.

DO YOU KNOW?

We live in a world where greed remains a part of the capital market and, unfortunately, where there is greed there is fraud. It can be reassuring to know that just a few steps can be taken to help detect suspicious activity and prevent a company from losing millions or even billions of dollars.

8.4.2. Acquiring floor Space and Plan space Layouts

Floor Space Layouts : Installation of a new system to replace the current system requires many facilities which are listed below :

- New Office
- Computer Rooms
- Production Layouts

The MIS project managers should perform the following operations :

- Prepare rough layouts of the MIS
- Estimate the floor area required
- Prepare a cost estimate and submit it to the management for approval

Plan Space Layouts : Space planning should be started as soon as the approval is obtained regarding the space allocations from the management. This should be done at an earliest mainly because of two reasons :

- There may be a long lead time if new partitions, electrical work, air conditioning or even new buildings are required.
- The detailed work flow depends upon the physical arrangements of the buildings.

Space planning includes :

- Space occupied by the people
- Space occupied by an equipment
- Movement of the people and equipment's in the work process
- Number of exists
- Kind of exists
- Location of utilities, outlets and controls
- Environmental requirement for an equipment

- Safety factors
- Working condition of the personnel

A large investment in good working conditions will repay its cost many times in the future.

8.4.3. Organizing the Implementation

As the implementation tasks have been defined, management usually assigns a project manager to guide the implementation process.

- In small companies, the person who is responsible for organizing the implementation process is the finance manager or computer center manager.

The role of line managers must be made very clear in this process. The main purpose of the management information system is to increase the quality and amount of their contributions. Top managers must involve middle managers in the organizing process. System specialists, computer programmers should also be involved.

Top management should make sure that each people who will operate the system should have active parts in the implementation.

8.4.4. Developing procedures for Implementation

After organizing the personnel for the implementation process, the next step in the implementation is developing procedures for implementation.

The project leader has the network plan for proceeding with the implementation process. The project leader must call the key people that are involved in the project to prepare more detailed procedures for the system installation.

For example : System analyst must develop procedures for delivering instructions and forms to the supervisor for a better coordination and integration.

- The procedures for evaluating and selecting hardware must also be developed.
- Procedures for operating the MIS in parallel must be developed.
- Testing of each segment of the total system should be done.

The testing procedures are :

- Which segment will be tested?
- When the testing will be done?
- Who will perform the testing operations?
- Who will evaluate the test results?
- Who will approve the system?
- Who will recommend modification?

8.4.5. Train the Operating Personnel

A program should be developed to impress the management and to train the operating personnel in their new duties.

Role of management :

- The people who participate in the development of the system come under the part of management.
- Here two short seminars are provided. These seminars are enough for these personnel's.
- The first meeting is arranged at the time of detailed designing. Some proposals which are valuable can be combined in the design.
- The second meeting is held at the end of the implementation stage. This meeting usually shows the benefits of the system and the role of the executives.

Train the first line supervisors :

- First line supervisors should have a proper understanding about the new MIS and its functions.
- They should know about the operations of the system and learn them to operate the system efficiently.

Role of supervisors : The supervisors should have the answers to the following questions :

- The new skills that are to be learned.
- Changes in procedure that are to be made.
- Measurement of performance.
- Jobs will either be upgraded or downgraded.

Professional Support Personnel :

- The professional support personnel are computer center personnel, marketing researchers and the accounting personnel.
- These personnel's provide input to the Management Information System. They are responsible for the processing of data and Information.
- The concerned personnel's are required to attend one or more orientation meetings.
- These people works with only a small part of the Management Information System. The seminars should be designed in such a way that they provide the personnel's with a complete understanding of the system.

8.4.6. Computer Related Acquisitions

The next step in the implementation process is to gather the computer acquisitions required by an organization. The computer related acquisitions can either be hardware or software.

(*a*) Hardware :

- In computer hardware, the design of the computer architecture available from the vendors and the system are closely tied together.
- A choice of CPU and peripheral devices are to be made.
- In this, it has to be decided whether the hardware is to be purchased or to be brought on lease.
- Capital analysis, usage etc. are the factors that are to be taken into consideration.

(*b*) Software :

- Many software packages are available commercially.
- Small companies must purchase the software products. But, the purchasing of the software has some problems. Sometimes the software requires so many modifications that it would be cheaper to develop the software internally.
- Large companies can either develop the software internally or under contract.

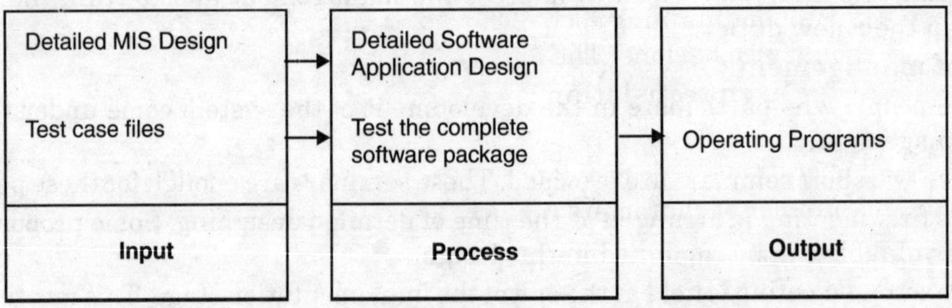

Fig. 8.3. Computer Related Acquisitions

(c) Personnel :

- It consists of a personnel planning chart. This chart shows the number of persons required in terms of skill, source etc. The source can either be internal or external.
- It provides an opportunity to upgrade the personnel after the training process.
- At higher levels, computers must be used to handle the jobs which were earlier performed by individuals manually.

(d) Materials :

- The two main materials that are required in a Management Information System are forms and manuals.
- Demand for the forms and manuals must be estimated so that it can be ordered.
- Computer materials like cards, tapes, printout paper should be ordered whenever required.

DO YOU KNOW?

Database management is expected to be one of the fastest growing occupations in 2012. As organizations continue to add and implement more and more technologies, jobs will continue to increase.

8.4.7. Develop the forms for Data Collection

The input of the Management Information System is the large amount of data. The data can either be internal or external to a company.

For example : In a sales estimation, the salesperson must sit in a room and fill the form summarizing the day's activities.

The major benefit of data collection through forms is :

- They ensure that right information is provided.
- It helps in simplifying the data processing that is required for computer storage.
- Forms are also required for sending data at intermediate stages.

8.4.8. Develop the Files

The next phase in the implementation process is the development of files. The actual data must be obtained and recorded for the initial testing and operation of the system.

The requirements of developing these files are :

- Format of data
- Storage form
- Storage format
- Remarks to show when the data has been stored

Procedures are required to be developed for updating each piece of data and the entire file.

(a) Master file : Master file is a file that collects the data that is used in various operations. A procedure must be developed to obtain the data from the environment. The responsibility of maintaining each file must be assigned to each person. The designing of forms should also be done. The form helps the programmer to analyze the data.

Requirement for each translation :

- File name
- Volume of operations
- Maximum number of characters required to record each element

8.4.9. Test the System

The testing of the system is done on the basis of the test specifications and procedures. Testing during the installation system requires the following :

- Component tests
- Subsystem tests
- System acceptance tests

In the component testing, each component of the system is tested. In the subsystem testing, subsystems of the organization are tested. Acceptance testing is performed on the entire system.

Components are :

- **Equipment :** These can either be old or new.
- New forms
- New software programs
- New procedures

Performance meters : Tests are performed for the following meters :

- Accuracy of the system
- Range of inputs
- Frequency of inputs
- Reliability

The main problem with the testing of the system is :

How the daily operations can be managed with the testing of the equipment's, forms, procedures etc.?

The solution to the problem is very simple.

Plan a different location for the new equipment that is far from the old equipment.

Requirements of system testing :

- Verification of multiple inputs
- Verification of logic system
- Timing aspect of the parts of the system
- Interface of the system

8.4.10. Cutover

Cutover is basically a point at which the new component replaces the old component or the new system replaces the old system.

Involvement for cutover operation :

- Physical transfer of files
- Movement of workstations
- Movement of people
- Arrangement of office furniture

In the cutover, the old forms and files are discarded. There are some problems faced in the cutover :

- System may have bugs despite the system and component testing. The main reason for the bugs in the system is due to improper training that is provided to the personnel. The persons who were earlier working with the old system are not comfortable with the new equipment's, procedures and co-workers.

The debugging process takes several months to complete.

8.4.11. Document the System

The documentation of the system is the last phase in the implementation process. The documentation of the system is the written description of the system. The parts that are documented are :

- Scope
- Purpose
- Procedures
- Information flow components

Why documentation is required?

Documentation is required for the :

- Replacements of the subsystems in an organization
- Interfacing with the other systems
- Training of the new operating personnel
- Evaluation of the system
- Updating the system

Benefits of documenting a system :

- A new team of operators could learn to operate the new system with the help of available documentation properly.
- When reconstruction of the new system is required, documentation is very helpful.
- Reference design is available for the managers and designers. These people are concerned with the system maintenance.
- It helps the analyst to gather valuable data for developing new Management Information System, schedules, plans and costs.

Documentation of manual MIS : The documentation of the manual MIS consists of :

- Summary of scope, types of outputs and user.
- Comparison between the kind of old and new people who are using the new system.
- Comparison between the number of old and new people who are using the new system.
- Flowcharts
- Forms
- Procedures
- Methods for controlling the system
- Methods for revising the system

Documentation of computer based MIS : In the documentation of computer based MIS, documentation of the following are performed :

- Software development
- Programs
- Files
- Input/output forms
- Codes
- Methods of entering data
- Methods of processing data
- Methods of retrieving data

Documentation of master file is also required.

Summary

➡ Implementation is an activity which starts when the system is developed and is ready to be installed in the organization.

- System analyst is the catalyst that is responsible for the entire change of process and makes sure that the changes are acceptable to all the persons that are working in an organization.
- The implementation of MIS can be achieved by using direct approach, parallel approach, modular approach and phase in approach.
- In the direct approach, the new system is installed directly with the immediate discontinuance of the old existing system.
- The direct approach is also referred to as "Cold Trunky" approach.
- In the parallel approach, the selected new system is installed and operated with the current system in parallel.
- In modular approach, the new system is introduced at one location only for the trial purpose.
- The modular approach is also known as "Pilot Approach".
- Phase in implementation deals with the segmentation of the system whereas the modular approach deals with the segmentation of an organization.
- Planning the implementation of MIS in an organization is the first step in the implementation phase. The planning process is a function of line management.
- A program should be developed to impress the management and to train the operating personnel in their new duties.
- The computer related acquisitions can either be hardware or software.
- The input of the Management Information System is the large amount of data. The data can either be internal or external to a company.
- Master file is a file that collects the data that is used in various operations. A procedure must be developed to obtain the data from the environment.
- The testing of the system is done on the basis of test specifications and procedures.
- In the component testing, each component of the system is tested.
- In the subsystem testing, subsystems of the organization are tested.
- Acceptance testing is performed on the entire system.
- Cutover is basically a point at which the new component replaces the old component or the new system replaces the old system.
- The documentation of the system is the last phase in the implementation process. The documentation of the system is the written description of the system.

Short Question Answers

1. Define Implementation of a Management Information System.

Ans. Implementation is an important managerial activity related to the adoption and management of an innovation. It is an activity which starts when the system is developed and is ready to be installed in the organization.

2. What are the factors that affect the outcome of the implementation process?

Ans. The factors that affect the implementation process are :

- Role of users in the implementation process.
- Amount of management support for the implementation process.
- Complexity of risk involved in the implementation activity.
- Level of risk involved in the implementation activity.
- Quality of management in the implementation process.

3. Who is a system analyst?

Ans. System analyst is the catalyst that is responsible for the entire change of process and makes sure that the changes are acceptable to all the persons that are working in an organization.

4. What are the approaches to achieve implementation?

Ans. Direct, parallel, modular and phase in are the four approaches to achieve implementation.

5. Explain direct approach.

Ans. In the direct approach, the new system is installed directly with the immediate discontinuance of the old existing system. It is also known as immediate cut over or change over.

6. Explain parallel approach.

Ans. In the parallel approach, the selected new system is installed and operated with the current system in parallel. It is an expensive approach.

7. Explain modular approach.

Ans. The modular approach is also known as "PILOT APPROACH". In this approach, the new system is introduced at one location only for the trial purpose. If its performance is according to the needs of an organization, it is introduced in the entire organization otherwise not. Thus, implementation of the MIS system in the organization is done on peace-meal basis.

8. Explain phase in approach.

Ans. The phase in implementation is quite similar to the modular approach. The only difference between the two is that; phase in implementation deals with the segmentation of the system whereas the modular approach deals with the segmentation of an organization.

9. What are the methods of Implementing MIS?

Ans. There are four methods of implementing a Management Information System. These are :

1. Install the new system in the new organization.
2. Cut off the old system and install a new system.
3. Cut over by segments.
4. Operate in parallel and cut over.

10. What are the Implementation Steps of a Management Information System?

Ans. The implementation steps are :

(*a*) Plan the implementation
(*b*) Acquire floor space and plan space Layouts
(*c*) Organize the Implementation
(*d*) Develop Procedures for Implementation
(*e*) Train the operating personnel
(*f*) Computer Related Acquisition
(*g*) Develops the forms for data collection
(*h*) Develop the files
(*i*) Test the system
(*j*) Cut over
(*k*) Document the system

11. What are the three phases in planning of the implementation process?

Ans. These three phases are :

- Initial Installation
- Test of the whole system
- Evaluation, maintenance and control of the system

12. What are the implementation tasks?

Ans. The major implementation tasks are :

- Planning the implementation activities
- Acquiring and laying out facilities and offices
- Organizing the personnel for implementation
- Developing procedures for installation and testing
- Developing the training program for operating personnel
- Completing the system's software
- Acquiring required hardware
- Generating files
- Designing forms
- Testing the entire system
- Complete cutover to the new system
- Documenting the system
- Evaluating the MIS
- Providing system maintenance

13. Explain the relationships among tasks.

Ans. The establishment of relationship among tasks is done for both the small and large projects.

- **Small Projects:** For the small projects, the order of performance may be described in a text format. A Gantt chart or network diagram makes visualization of the plan and schedule much clearer.
- **Large Projects:** For large projects, many parallel and sequential activities are interrelated so that a network diagram can also be used.

14. How is MIS established?

Ans. Management Information System should be established because of the following reasons :

- Monitoring the progress of Implementation
- Proper control of activities
- Efficient Information system should be developed.

15. What does space planning includes?

Ans. Space planning includes :

- Space occupied by the people
- Space occupied by an equipment
- Movement of the people and equipment's in the work process
- Number of exists
- Kind of exists
- Location of utilities, outlets and controls
- Environmental requirement for an equipment
- Safety factors
- Working condition of the personnel

16. What procedures are developed for the implementation process?

Ans. The procedures that are developed for the implementation process are :

- Procedures for evaluating and selecting hardware must also be developed.
- Procedures for operating the MIS in parallel must be developed.
- Testing of each segment of the total system should be done.

17. What is a master file?

Ans. Master file is a file that collects the data that is used in various operations. A procedure must be developed to obtain the data from the environment.

18. What are the requirements for developing the files?

Ans. The requirements for developing the files are :

- Format of data
- Storage form
- Storage format
- Remarks to show when the data has been stored.

19. What is the benefit of data collection through forms?

Ans. The major benefit of data collection through forms is :

- They ensure that right information is provided.
- It helps in simplifying the data processing that is required for computer storage.
- Forms are also required for sending data at intermediate stages.

20. Why documentation is required?

Ans. Documentation is required for the :

- Replacements of the subsystems in an organization
- Interfacing with the other systems
- Training of the new operating personnel
- Evaluation of the system
- Updating the system

Review Questions

1. Explain the approaches to achieve Implementation in detail.
2. What are the methods of Implementing MIS? Explain.
3. Explain the implementation steps of a Management Information System in detail.

□□□

Chapter 9

Evaluation, Control and Maintenance of Management Information Systems

"Everybody gets so much information all day long that they lose their common sense."

Inside this Chapter

- Evaluate the Management Information System
- Concepts of the Evaluation Process
- Structure of the Evaluation Process
- Control the Management Information System
- Maintain the Management Information System
- Environmental Change
- Internal Problems in System Maintenance
- System Maintenance Responsibility
- Maintenance Projects
- System Planning

9.1. EVALUATE THE MANAGEMENT INFORMATION SYSTEM

The evaluation of the Management Information System is done after a small period of time. The evaluation is performed on each step of the :

- Design
- Final system performance

In the evaluation of a system, principle of feedback is applied. The process of evaluation should be performed as early as possible because if it is delayed, it becomes difficult for the designer to remember each and every important detail.

Key People : The key people involved in the evaluation of a management information system are :

1. Customers
2. Designers
3. Financial Specialists

The evaluation is performed by both the designers and customers. The financial specialists should evaluate the project in the following terms :

- Planned cost vs. actual cost of the design of the system
- Implementation
- Operation

9.2. CONCEPTS OF THE EVALUATION PROCESS

The concepts that are involved in the evaluation process are :

1. **System Integrity :** The subsystems are integrated into the total system without any redundancy. It provides the following advantages :
   - Flexibility
   - Expandability
2. **Operation Integrity :** The operation integrity involves skilled people that operate the system. A backup should be available that should prevent the system in case of a system failure.
3. **Internal Integrity :** The internal integrity provides the following factors :
   - The validity of the output of the system.
   - Security of the system against human error, manipulation or theft.
4. **Procedural Integrity :** The procedural integrity involves :
   - Proper documentation of the system and procedures.

9.3. STRUCTURE OF THE EVALUATION PROCESS

The structure of the evaluation process is measured in terms of :

Measurement of cost or benefit = Measurement of Old – Measurement of New

The measurement of change is related to the basic goals of a management information system. It is the responsibility of the designer to select the level at which the measurement process should take place.

When the proper variables of interest are identified, a table should be set to formalize the measurement process.

DO YOU KNOW?

The vision is really that in the information age the micro-processor-based machine, the PC, along with great software, can become sort of the ultimate tool dealing with not just text, but numbers and pictures, and eventually, even difficult things like motion video.

–Bill Gates

9.4. CONTROL THE MANAGEMENT INFORMATION SYSTEM

The control of the Management Information System is done by the line managers. The control of the system means that the operations should operate as it were designed to operate. The operators must develop their own procedures or may modify the procedures that were already designed.

9.5. MAINTAIN THE MANAGEMENT INFORMATION SYSTEM

The maintenance of the management information system is very similar to the controlling of the system. **Maintenance** is basically an activity that helps to keep the MIS system at the highest level of effectiveness and efficiency within the cost limitations. It helps in reducing the errors that are caused by the :

- Designers
- Environmental Change

The maintenance process is applied to the following activities :

- Change in the forms
- Change in the operating system
- Change in the procedures that are used

- Modification in the hardware
- Modification in the software
- Changes in the policy statements

The maintenance activities are classified into three types. These are :

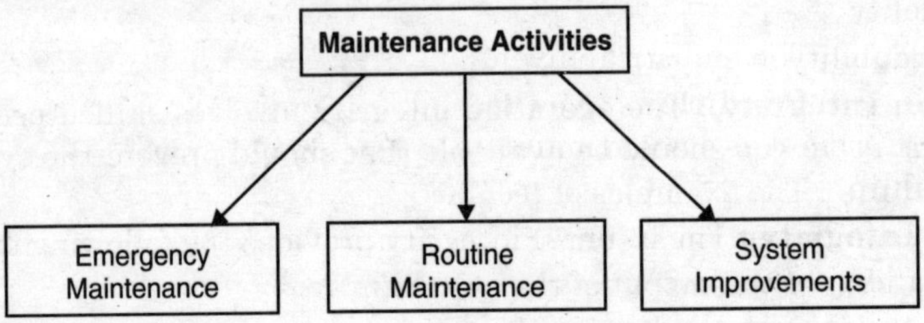

Fig. 9.1. Maintenance Activities

9.6. ENVIRONMENTAL CHANGE

The environmental change includes four factors. These are described below :

(a) Governmental Policies : Specialists are required in many companies. The role of these specialists is to remain updated about the various changes that take place in the requirement process. The governmental policies, regulations and legislations keep on changing from the government side. These factors require constant updating of the MIS in a company.

For Example : Banks must be aware of the :

- New regulations
- Interest rate
- Minimum down payments that can be made

(b) Economic Conditions : The economic conditions play an important role in defining the financial information of the system. The system should be properly defined. It should meet the needs of the :

- Users
- Accounting department
- Finance department

For Example : Any change in the unemployment rate can affect the :

- Labour cost
- Time required to complete a project

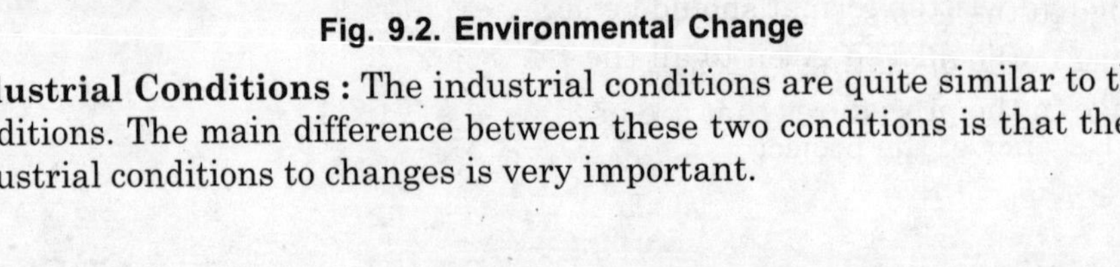

Fig. 9.2. Environmental Change

(c) Industrial Conditions : The industrial conditions are quite similar to the economic conditions. The main difference between these two conditions is that the reaction of industrial conditions to changes is very important.

The areas that are affected by the industrial conditions are :

- Price
- Policy
- Hiring methods
- Budget

For Example : The industrial inventions that are used to gather and report the data.

(*d*) **Technological Conditions :** New technology is created and is used by a large number of people. Continuous system maintenance is required for this factor.

For Example : The examples of technological factors are :

- Data communication systems
- High storage systems

9.7. INTERNAL PROBLEMS IN SYSTEM MAINTENANCE

There are many problems that are related to the maintenance of the management information system. These are :

- No proper plan
- No proper allocation of resources
- Lack of management understanding problems
- Improper documentation
- Lack of understanding

9.8. SYSTEM MAINTENANCE RESPONSIBILITY

The maintenance is done by specific people. These are :

- Specific responsibility is given to the supervisor.
- Special team of MIS analyst, specialist and programmer is formed to perform system maintenance.

If the tasks are not assigned to special people, problems may arise.

9.9. MAINTENANCE PROJECTS

Maintenance projects are created to form :

- Error reports
- Request changes by the users
- Member of the maintenance team
- Management of the company

9.10. SYSTEM PLANNING

A proper system planning is required to perform the maintenance of the management information system in use. Maintenance cannot be performed on a first come first serve basis.

The factors that must be taken into account are :

- Record all the requests that are made for the change process. Only requests that are made in a written format should be accepted and recorded.
- A priority should be given to all the requests. The priority is made on the following factors :
  - Urgency of the project

 – Long term benefits provided by the project
 – Time required
 – Resources required
- Make annual and monthly plans.
- Document the maintenance process.

Summary

➡ The process of evaluation should be performed as early as possible because if it is delayed, it becomes difficult for the designer to remember each and every important detail.

➡ The key people involved in the evaluation of an information system are customers, designers and financial specialists.

➡ The concepts that are involved in the evaluation process are system, operation, internal and procedural integrity.

➡ The subsystems are integrated into the total system without any redundancy.

➡ The operation integrity involves skilled people that operate the system.

➡ The procedural integrity involves proper documentation of the system and procedures.

➡ The measurement of change is related to the basic goals of a management information system. It is the responsibility of the designer to select the level at which the measurement process should take place.

➡ The control of the Management Information System is done by the line managers.

➡ The control of the system means that the operations should operate as it were designed to operate.

➡ Maintenance is basically an activity that helps to keep the MIS system at the highest level of effectiveness and efficiency within the cost limitations.

➡ Emergency, routine and system improvement are the three maintenance activities.

➡ Specialists are required in many companies. The role of these specialists is to remain updated about the various changes that take place in the requirement process.

➡ The economic conditions play an important role in defining the financial information of the system.

➡ A proper system planning is required to perform the maintenance of the management information system in use. Maintenance cannot be performed on a first come first serve basis.

Short Question Answers

1. What is the evaluation of the Management Information System?

Ans. The evaluation of the Management Information System is done after a small period of time. The evaluation is performed on each step of the design and final system performance.

2. Who are the key people involved in the evaluation process?

Ans. The key people involved in the evaluation of a management information system are customers, designers and financial specialists.

3. What are the terms at which the specialists should evaluate the system?

Ans. The financial specialists should evaluate the project in the following terms :

- Planned cost vs. actual cost of the design of the system

- Implementation
- Operation

4. What are the concepts involved in the evaluation process?

Ans. The concepts that are involved in the evaluation process are system, operation, internal and procedural integrity.

5. Explain system integrity.

Ans. The subsystems are integrated into the total system without any redundancy. It provides flexibility and expandability.

6. Explain operation Integrity.

Ans. The operation integrity involves skilled people that operate the system. A backup should be available that should prevent the system in case of a system failure.

7. Explain internal integrity.

Ans. The internal integrity provides the following factors :

- The validity of the output of the system.
- Security of the system against human error, manipulation or theft.

8. Explain procedural integrity.

Ans. The procedural integrity involves proper documentation of the system and procedures.

9. Explain the structure of the evaluation process.

Ans. The structure of the evaluation process is measured in terms of :

Measurement of cost or benefit = Measurement of Old – Measurement of New

The measurement of change is related to the basic goals of a management information system. It is the responsibility of the designer to select the level at which the measurement process should take place.

10. Explain the controlling of management information system.

Ans. The control of the Management Information System is done by the line managers. The control of the system means that the operations should operate as it were designed to operate.

11. Explain the maintenance of management information system.

Ans. Maintenance is basically an activity that helps to keep the MIS system at the highest level of effectiveness and efficiency within the cost limitations. It helps in reducing the errors that are caused by the designers and environmental change.

12. What are the activities on which the maintenance process is applied?

Ans. The maintenance process is applied to the following activities :

- Change in the forms
- Change in the operating system
- Change in the procedures that are used
- Modification in the hardware
- Modification in the software
- Changes in the policy statements

13. State the classification of maintenance activities.

Ans. The maintenance activities are classified into three types. These are emergency maintenance, routine maintenance and system improvement.

14. What are the environmental change factors?

Ans. There are four environmental factors. These are governmental policies, technological factors, economic and industrial conditions.

15. Explain governmental policies.

Ans. Specialists are required in many companies. The role of these specialists is to remain updated about the various changes that take place in the requirement process. The governmental policies, regulations and legislations keep on changing from the government side. These factors require constant updating of the MIS in a company.

16. Explain the economic conditions.

Ans. The economic conditions play an important role in defining the financial information of the system. The system should be properly defined. It should meet the needs of the users, accounting and finance department.

17. Explain the industrial conditions.

Ans. The industrial conditions are quite similar to the economic conditions. The main difference between these two conditions is that the industrial conditions reaction to changes is very important. The areas that are affected by the industrial conditions are price, policy, hiring methods and budget.

18. Explain technological conditions.

Ans. New technology is created and is used by a large number of people. Continuous system maintenance is required for this factor. The examples of technological factors are data communication systems and high storage systems.

19. What are the internal problems faced in the system maintenance process?

Ans. There are many problems that are related to the maintenance of the management information system. These are :

- No proper plan
- No proper allocation of resources
- Lack of management understanding problems
- Improper documentation
- Lack of understanding

20. Explain system maintenance responsibility.

Ans. The maintenance is done by specific people. These are :

- Specific responsibility is given to the supervisor.
- Special team of MIS analyst, specialist and programmer is formed to perform system maintenance.

21. Explain system planning.

Ans. A proper system planning is required to perform the maintenance of the management information system in use. Maintenance cannot be performed on a first come first serve basis.

Review Questions

1. Explain the evaluation process in detail.
2. Explain the environmental change.
3. Explain the control and maintenance of the system.
4. Explain system planning in detail.

□□□

Chapter 10

Pitfalls in Management Information System

" Any sufficiently advanced technology is indistinguishable from magic."

Inside this Chapter

- Types of Pitfalls

Management Information System are risky systems since they aim at changing the organizations. The main problems faced by an organization are :

- New group of stakeholders
- New technology
- Less experienced professionals
- Partly implemented functions
- Over budget
- Delay over implementation

These problems results in losses to an organization and improper use of resources.

10.1. TYPES OF PITFALLS

There are four pitfalls in a management information system. These are :

(A) Fundamental Weaknesses

Fundamental weaknesses are faced in an organization only. When a company decides to develop a Management Information System, these fundamental weaknesses are faced.

Review

I. **No Management System to build Upon :** The MIS must be built on the top management system. This includes :

- Organizational arrangement
- Structure for proper planning and control

- Procedure for proper planning and control
- Clear creation of objectives

The process of management is not performed well due to the following reasons :

- Lack of managerial applications
- Lack of operational applications

Role of manager : The role of manager is :

- Establishment of management system
- Repairing of the management system
- Modify the management system

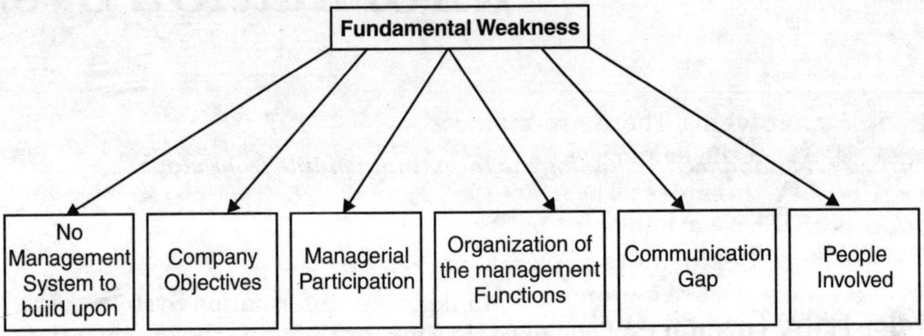

Fig. 10.1. Fundamental Weakness

II. Company Objectives : The Company without objectives is like a company without a statement of mission and purpose. This lacking of the statement of mission and purpose is a common weakness. This is the weakness caused by the top management of the company due to which the entire employee workforce is left without the benefit of the valuable guidance that the objectives provide. The major benefits of setting up of clear objectives are :

- Helps in decision making
- Helps in inference purposes

Suppose if it is not clear in what business we are actually, then the major challenge the organization faces is entirely different and should be analyzed from the ground level. If there would be a mission statement, some of the problems could be handled easily.

Also if there will be no company objectives, then there are fewer chances that the MIS will satisfy all the needs of an organization.

III. Managerial Participation : The development of the Management Information System is the responsibility of the management. Managerial participation helps in the development of a better and effective information system. The management includes :

- Top level management
- Operating line management

For Example : Top companies like Ford, IBM and Microsoft etc. are successful because in these companies both the management and users are required in the designing phase.

IV. Organization of the MIS Functions : The lack of organization of the MIS functions leads to the system failure.

V. Communication Gap : In an organization, communication gap is faced at various levels. The computer technician will never speak the language of the management and vice versa. Thus due to the communication gap, design problem arises.

Stages in MIS development	Communication gap
Designer asks the information form the user.	User does not provide the designer with proper information.
A plan is created by the designer and is provided to the programmer.	Designer converts the data heard from the user in the form of charts. This leads to data alteration.
Programmer implements the system.	Programmer includes his own ideas and thus further alters user needs. The result obtained is not as per the needs of the user.

VI. People Involved : There are many key people involved in an organization. The performance of the employees can be classified as :

- **Good Performers :** These are those people in an organization that provide good results. These are the best performers.
- **Average Performers :** Average Performers are the people that are productive only in the right environment. These are better than the poor performers.
- **Poor Performers :** Poor performers produce negative results and some other person has to be appointed to fix up their messes.

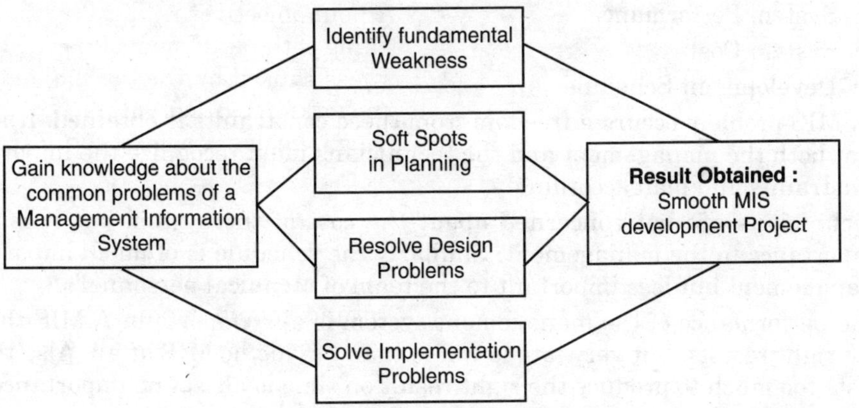

Fig. 10.2. Pitfalls in MIS

(B) Soft Spots in Planning

Soft spots in planning are related to the Management Information System that is under planning and development.

Review

I. MIS response to a Business Plan : The main problems that are faced by the MIS planning are that the personnel's associated with the planning process forgets the original basic purpose of the management. The professionals face difficulties for responding just to the information needs of the main business. One of the difficulties is the temptation to purchase the latest hardware and software techniques. When a change is made to the business plan, the corresponding MIS must be made and changed.

Each MIS plan must be a proper response to a business plan.

DO YOU KNOW?

Safety Information management system are used in environments that emphasize the safety of employees and proper use of resources to avoid accidents.

II. **One for one conversion :** The one for one conversion provides an opportunity to upgrade an existing system into a Management Information System. Suppose a technician converts an existing computer system or a manual and converts it without upgrading or changing it.

III. **System View :** In a system view, there should be a master plan available with the help of which MIS planning and development can be performed. Without a master plan, a system failure occurs.

IV. **Setting Project and System Objectives :** System objectives are required for the development of a correct plan. Also, objectives cannot be made until some of the general planning variables are known. Thus both the setting up of planning and objectives goes hand in hand. Following parameters should be taken into consideration :

- What kind of system is required?
- When is the system required?
- What is the cost of the system?

By using the above mentioned parameters, both the objectives and the plan can be achieved in a synchronized manner.

V. **Facing Constraints :** Constraints are faced on the following :

- Financial Definition
- System Performance
- System Cost
- Development Schedule

An MIS problem occurs if freedom from these constraints is obtained. It is important that both the management and the technicians must recognize the importance of the constraints and plan accordingly.

Technicians are not concerned about the cost of the system but it is of supreme importance to the management. Similarly the schedule is of much importance to the management but less important to the team of technical personnel's.

The performance of the management system is also important. A MIS that produces the right results, but very late is worse than having no MIS at all. Also the MIS that costs too much to produce the right result on time is also of no importance.

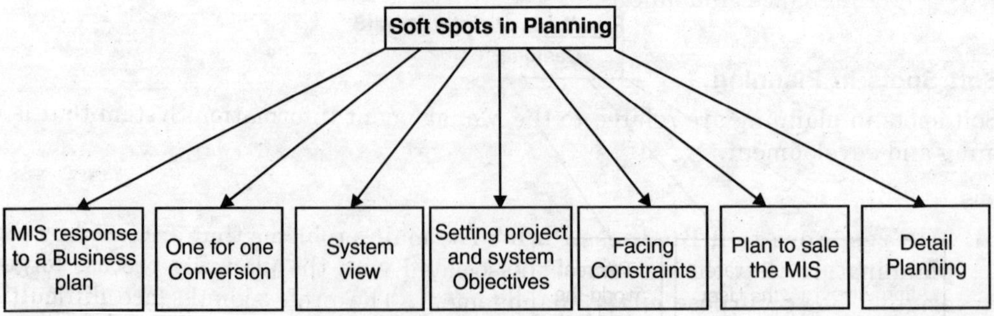

Fig. 10.3. Soft Spots in Planning

VI. Plan to Sale the MIS : System designers face the problem of gaining acceptance from the user for whom the system is designed. Thus it should be taken care that the system should be designed in such a way that it is easily accepted by the users. It is the toughest part of the entire designing and implementation phase.

VII. Detail Planning : The MIS developers must plan in detail each phase of the project.

(C) Design Problems

Design problems are the pitfalls faced during the designing phase. These pitfalls are under the control of managers and can be easily solved.

Review

I. Consider Alternative Designs : During the early phase of the designing process, many ideas should be considered. It is the responsibility of the designer to provide the various alternative designs with the positive and negative impact of each. As the time passes, the frequency of alternative designs also decreases.

II. Beware the User Interface : Beware the User Interface is partially a technical problem. The user interface should be according to a user. User should be comfortable with the design or interface of the system. The advantage is that it helps in the conversation among the computer experts and eliminates the conversion between the technician and user.

III. The Real World-The Acid Test : The business organizations are not research institutes. Thus, the MIS is being implemented to support the firm's main line of business and not to extend the state of art in MIS design.

IV. If it Moves-Automate it : Only some things can be automated and not all.

V. The Computer Obsession : Computer should not be an obsession in a system. It is a tool and should be used as a tool only.

VI. Documentation : Documentation is a problem that is faced in the planning, designing and implementation phase. The following items must be documented in detail :

- Plans
- Project and system objectives
- Specification of functions
- Specification of performance
- User interface specifications
- User instruction manual
- Reference manual
- Maintenance guideline.

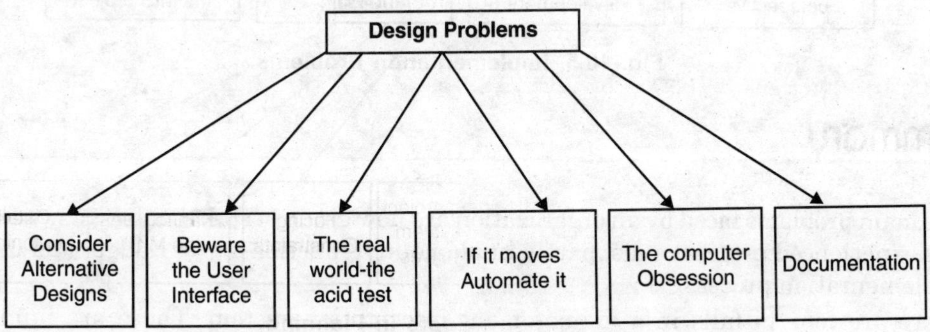

Fig. 10.4. Design Problems

These items are necessary to be managed. Not documenting these things in detail is a guarantee of failure in some part of the operation of MIS.

(D) Implementation Problems

Various types of implementation problems are faced in the development of a management information system.

Review

I. **Hardware Perspective :** In the hardware perspective, the managers show off their computer centers, disk space and the millions of instructions the processor executes per second. Nowadays the hardware items are becoming cheap and are of little concern in the MIS purchase design.

II. **Software Development :** The software development pitfall includes two pitfalls. These are :
- Hardware related
- Personnel related

The hardware performs the instructions. Therefore wrong instructions lead to wrong results.

III. **Test it and Test It Again :** A major error that is seen is improper planning that is done to perform the testing. The rule that is used in project planning is :

1/3 planning and design, 1/3 implementation, 1/3 testing

Testing must be performed at the :
- Functional level
- Component level
- System level

Problems are experienced at each level and they must be fixed.

IV. **Controlling the MIS Project :** Controlling the MIS Project is the one of the four basic management functions. It also remains one of the major causes of MIS development project failure.

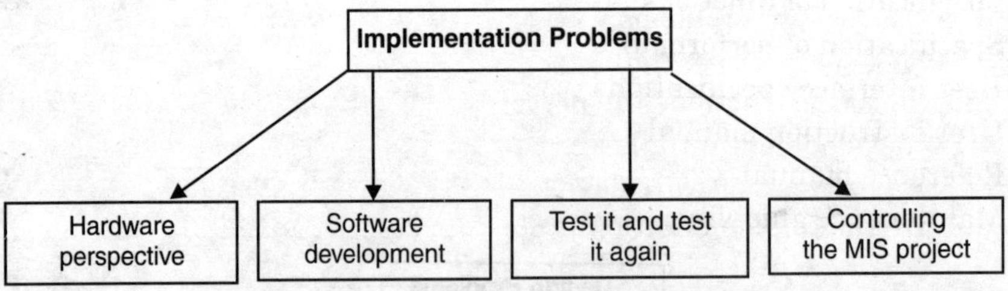

Fig. 10.5. Implementation Problems

Summary

- The main problems faced by an organization are new group of stakeholders, new technology, less experienced professionals, partly implemented functions, over budget and delay over implementation process.
- There are four pitfalls in a management information system. These are fundamental weakness, soft spots in planning, design problems and implementation problems.

- The role of manager is establishment of management system, repairing of the management system and modification of the management system.
- No management system to build upon, company objectives, organization of the management functions, communication gap, managerial participation and people involved are the fundamental weaknesses.
- The benefits of setting up of clear objectives are decision making and helps in inference purposes.
- Managerial participation helps in the development of a better and effective information system.
- The performance of the employees can be classified as good performers, average and poor performers.
- The main problems that are faced by the MIS planning are that the personnel's associated with the planning process forgets the original basic purpose of the management.
- The one for one conversion provides an opportunity to upgrade an existing system into a Management Information System.
- In a system view, there should be a master plan available with the help of which MIS planning and development can be performed. Without a master plan, a system failure occurs.
- System objectives are required for the development of a correct plan. Also, objectives cannot be made until some of the general planning variables are known.
- MIS response to a business plan, one for one conversion, setting project and system objectives, system view, plan to sale the MIS, facing constraints and detail planning are the soft spots in planning.
- Beware the User Interface is partially a technical problem. The user interface should be according to a user.
- Documentation is a problem that is faced in the planning, designing and implementation phase.
- Plans, project and system objective, specification of functions, specification of performance, user interface specifications, user instruction manual, reference manual and maintenance guideline are the design problems.
- In the hardware perspective, the managers show off their computer centres, disk space and the millions of instructions the processor executes per second.
- The software development pitfalls are hardware and personnel related pitfalls.
- Hardware perspectives, software development, test it and test it again and controlling the MIS project are the implementation problems.

Short Question Answers

1. What are the problems that cause loss to an organization?

Ans. The main problems faced by an organization are new group of stakeholders, new technology, less experienced professionals, partly implemented functions, over budget and delay over implementation process.

2. What are the four pitfalls in a management information system?

Ans. The four pitfalls in a management information system are fundamental weakness, soft spots in planning, design problems and implementation problems.

3. What are the fundamental weaknesses?

Ans. No management system to build upon, company objectives, organization of the management functions, communication gap, managerial participation and people involved are the fundamental weaknesses.

4. What are the benefits of setting up company objectives?

Ans. The major benefits of setting up of clear objectives are that it helps in decision making and inference purposes. Also if there will be no company objectives, then there are fewer chances that the MIS will satisfy all the needs of an organization.

5. What do you mean by communication gap faced in an organization?

Ans. In an organization, communication gap is faced at various levels. The computer technician will never speak the language of the management and vice versa. Thus due to the communication gap, design problems arises.

6. What is the benefit for one for one conversion?

Ans. The one for one conversion provides an opportunity to upgrade an existing system into a Management Information System. Suppose a technician converts an existing computer system or a manual and converts it without upgrading or changing it.

7. Why is system objectives required?

Ans. System objectives are required for the development of a correct plan. Also, objectives cannot be made until some of the general planning variables are known. Thus both the setting up of planning and objectives goes hand in hand.

8. Where are the constraints faced?

Ans. Constraints are faced upon financial definition, system performance, system cost and development schedule.

9. What is the problem faced while selling the MIS?

Ans. System designers face the problem of gaining acceptance from the user for whom the system is designed. Thus it should be taken care that the system should be deigned in such a way that it is easily accepted by the users. It is the toughest part of the entire designing and implementation phase.

10. What are the items that must be documented?

Ans. Plans, project and system objective, specification of functions, specification of performance, user interface specifications, user instruction manual, reference manual and maintenance guideline are the items that must be documented.

11. What are the levels at which the testing should be performed?

Ans. Testing must be performed at the functional level, component level and system level.

12. What are the soft spots in planning?

Ans. MIS response to a business plan, one for one conversion, setting project and system objectives, system view, plan to sale the MIS, facing constraints and detail planning are the soft spots in planning.

13. What are the design problems?

Ans. Plans, project and system objective, specification of functions, specification of performance, user interface specifications, user instruction manual, reference manual and maintenance guideline are the design problems.

14. What are the implementation problems?

Ans. Hardware perspectives, software development, test it and test it again and controlling the MIS project are the implementation problems.

15. What are the two pitfalls in software development?

Ans. The software development pitfall includes two pitfalls. These are hardware related and personnel related.

Review Questions

1. Explain fundamental weaknesses in detail.
2. Explain soft spots in planning in detail.
3. Explain design problems in detail.
4. Explain implementation problems in detail.

❑❑❑

Chapter 11

Enterprise Resource Planning

"A common mistake that people make when trying to design something completely foolproof was to underestimate the imagination of complete fools."

Inside this Chapter

- Enterprise Resource Planning Overview
- Evolution of ERP Systems
- Features and purpose of an ERP
- Reasons for Implementing ERP
- ERP Selection Process
- ERP System Architecture
- Components of an ERP
- Implementation of an ERP
- ERP Lifecycle
- ERP Key Vendors
- Commercial Applications
- Advantages of an ERP system
- Disadvantages of an ERP system

11.1. ENTERPRISE RESOURCE PLANNING OVERVIEW

The information in large organizations is spread across various computer systems. Each information subsystem is capable of handling a specific business activity but the enterprise's total performance is hindered by the lack of integrated information. With the arrival of E-business and the need to control multiple sources of Information within the enterprise, ERP software emerged as a major area of interest for many business organizations. It simplifies information sharing and integrates different functions so that the enterprise can be operated more effectively and efficiently. It makes use of a unified data store and consistent processes.

Definition : Enterprise resource planning is an integrated computer based application that is used to manage internal and external resources, financial resources, materials and human resources.

American Production and Inventory Control Society (2001) has defined ERP systems as a: "Method for an effective planning and controlling of all the resources needed to take, make, ship and account for customer orders in a manufacturing, distribution or service company."

They include modules which support functional areas like :

(*a*) Planning

(*b*) Manufacturing

(c) Sales
(d) Marketing
(e) Distribution
(f) Accounting
(g) Project management
(h) Inventory management
(i) E-business
(j) Transportation
(k) Service and maintenance

11.2. EVOLUTION OF ERP SYSTEMS

ERP system evolved with the remarkable developments in the field of hardware and software systems. During the era of1960s, most of the organizations designed, developed and implemented centralized computing systems. They automated their inventory control systems with the help of inventory control packages (IC).

Material requirements planning (MRP) systems were developed in the 1970s. It involved planning the product or parts requirements according to the master production schedule. New software systems known as **manufacturing resources planning (MRP II)** were introduced in the 1980s with an emphasis on optimizing manufacturing processes by synchronizing the materials with production requirements. MRP II included areas such as shop floor and distribution management, project management, finance, human resource and engineering.

ERP systems first appeared in the late 1980s and in the beginning of the 1990s with an additional feature of inter-functional coordination and integration. During the 1990s, ERP vendors added more modules and functions as "add-ons" to the core modules giving birth to the "extended ERPs." These ERP extensions include :

- Advanced planning and scheduling (APS)
- E-business solutions such as customer relationship management (CRM) and supply chain management (SCM).

11.3. FEATURES AND PURPOSE OF AN ERP

Features : The features of an ERP are listed below :

- It is built on a centralized database.
- It makes use of a common computing platform.
- It can reside on a centralized server or can be distributed across a modular software or hardware unit that provides services and communicates on a LAN.
- **Distributed :** It allows business to assemble modules from different vendors without the need for fixing of multiple copies of complex and expensive computer systems.

Purpose of an ERP : The main purpose of an ERP is to facilitate the flow of information between all business functions inside the boundaries of an organization. Other purposes are :

- Improve productivity
- Enable easy communication with clients and partners
- Integration of systems
- Provide flexibility
- It helps in making quick and vital decision making.

- Improving the return on investment.

The architecture of the software includes transparent integration of modules, providing flow of information between all the functions within an enterprise in a consistently visible manner. Corporate computing with ERPs allows companies to implement a single integrated system by replacing their incompatible legacy information systems.

DO YOU KNOW?

If an ERP project becomes too large, it will collapse into a black hole. Time and money are absorbed but nothing ever comes out.

11.4. REASONS FOR IMPLEMENTING ERP

Some of the top reasons for implementing ERP are :

1. **Integration of systems :** Integrated business systems helps the department to work more efficiently and effectively. It helps the organization to improve production cycles. It helps in managing the work processes and gets an easy view of the work flow.
2. **Enables new business to grow :** It helps a growing organization to handle number of tasks efficiently through ERP software. Understanding of the complete work flow enables an organization to remove the loopholes in functioning.
3. Internet and wireless networks helps the organization to grow and publicize their products. They also help to stay in touch with customers and business partners.
4. **Reduce expense on business processes :** The enterprise system helps the organization to work efficiently which helps in reducing the expenses on business processes.
5. **Effective decision making :** The management can take decisions with the help of previous data of the company and online services.
6. **Technological advancement for extending business services :** The business services can be extended with the help of technological advancements. Internet plays an important role in this. With the help of Internet, business processes can be promoted online rather than spending large amount of money by marketing manually.
7. **Improve communication with partners and customers :** ERP system enables an organization to send emails and messages to the customer without manual intervention. It also helps to provide easy customer service. The ERP software helps the company to retain the client and increase the client base.
8. **Easily manage the Human Resource :** ERP system helps in easily managing the process of shifts, schedules and attendance.

The software package should have the following characteristics so that it can be considered as an ERP system :

- It should be integrated.
- Should operate in real time.
- All applications should access one database to reduce redundancy.
- All the modules should have the same look and feel.

11.5. ERP SELECTION PROCESS

The most important selection criteria are the modules that are offered by an ERP. This also varies according to the needs of an organization. Following are the selection criteria of an ERP process :

1. **Customization :** Need of the software varies with each organization. Thus the companies need to adapt to the available software in the market according to their own use. But, customization should not cause any problem in updating the software's future releases.
2. **Implement-ability :** Each organization has different requirements. Thus it becomes very important to choose an ERP system that is implementable.
3. **Maintenance**: The ERP software should support multi-company, multi division and multi-currency environments. These three supports are important because there should not be any limitations in updating the system whenever an additional feature is available.
4. **Real Time Changes :** The ERP modules should work in real time with online and batch processing features. This helps in providing same consistent available information to all the departments of an organization. Further no errors are reported in case the system is not up to date.
5. **Flexibility** : This denotes the capability of the system in supporting the needs of an organization over its lifetime. As the business requirements of the organization change, one should be able to add extra modules. The ERP should be flexible enough to suit the organizational culture and business strategy.
6. **User Friendliness :** The end-users of an ERP system are the people who are not computer experts. Thus the products should not be too complex for an end user because the efficiency of end users directly affects the efficiency of an organization.
7. **Cost :** The cost of an ERP system is an important issue. The implementing organization may be a small or medium sized enterprise and they may not be as comfortable as a large, multi-national organization. ERP systems are generally complex systems that involve high costs.
8. **Systems Requirements :** It is important to choose an ERP that is independent of hardware, operating system and the database system. These system requirements determine the longevity of the product.
9. **After Sales Support & Training :** The ERP systems are complex applications so it cannot be learned by oneself. Thus, it is the responsibility of the vendor to provide training, sales and support. Every department should have their own software and a specialized kind of training is provided to each employee of the department for the efficient use of software.
10. **Back-up System :** To maintain the security of the systems, the backup unit of the system should be reliable. Users should be able to schedule routine and partly back-ups. The backup unit should also offer a solution for restoring the system within the shortest time.
11. **Reporting & Analysis Features :** The management team should be able to implement their own reporting and analysis tools.
12. **Vendor Credentials :** The commitment of the vendor to the product is known through the vendor's market share, reputation, number of consultants, number of installations performed, infrastructure support and demonstration of previous implementations.
13. **Integration with Other Software :** The modules should be integrated with other modules so that the whole data can flow among other modules. This leads to an increase in operational transparency. The data import and export are the widely used techniques.
14. **Internet Integration :** Sometimes the ERP system does not have built in modules like E-business, E-commerce etc. In this case, Internet integration should be available.

15. **Financing Options :** Financing options is an important criteria for an organization to know how to pay the investment and for how long to pay it.

11.6. ERP SYSTEM ARCHITECTURE

The architecture of an ERP system is discussed below :

- The modules of an ERP system can either work as separate units or several modules can be combined together to form an integrated system.
- The systems are usually designed to operate under several operating platforms such as UNIX, MS Windows NT, and Windows 2000 etc.

An Enterprise system makes use of two technologies :

- Thin client/server (C/S) technology
- Client/fat server (C/FS) architecture

Thin client/server (C/S) technology : In a C/S system, a number of client devices are operated by end users. They request services from application servers, which in turn get the requested service-related information from the database servers. The requests can be the following :

- Simple data files
- Data values
- Communication services
- Transaction processing or master file updates.

Three-tier architecture : In three tier architecture, the user interface runs on the client. Here, powerful PC's and servers are required to perform numerous operations. The three layers of logic in this architecture are :

1. **Presentation Layer :** It consists of Graphical user interface (GUI), browser for data entry and accessing of system functions.
2. **Application Layer :** It consists of business rules, functions, logic and programs which act on data that is received or transferred from and to the database servers.
3. **Database Layer :** It is the management of the organization's operational or transactional data including the metadata. It mostly employs industry standard RDBMS with structured query language (SQL) provisions.

This logical arrangement helps to perform the following functions :

- It helps the ERP user interface to run on the clients.
- Allows the processing modules to run on the middle-tier application servers.
- It helps the database system to run on the database servers.

DO YOU KNOW?

The ERP implementation isn't finished until the last user is dead.

11.7. COMPONENTS OF AN ERP

The components of an ERP system are :

1. **Transactional Backbone :** The transactional backbone includes financial, distribution, human resources and product life cycle management.
2. **Advanced Applications :** The advanced applications include customer relationship management, supply chain management and warehouse management. The supply chain management can further be classified into :
   - Purchasing
   - Manufacturing

- Distribution

3. **Management Portal :** The management portal comprises of the Decision Support System.

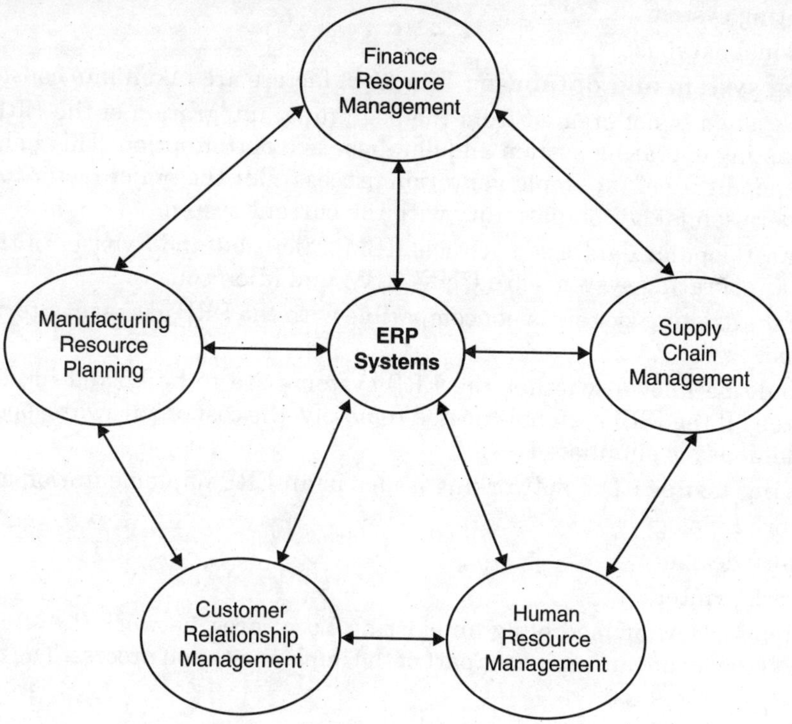

Fig. 11.1. ERP components

11.8. IMPLEMENTATION OF AN ERP

The implementation of ERP software is very complex for in-house developers. The main reason for this is the lack of desired skills. So it is advised to hire professionally trained people to implement the ERP systems.

The three types of services that must be taken care of are :

- Consulting
- Customization
- Support

The length and time to implement an ERP system depends on the following factors :

- Size of the business
- Number of modules
- Scope of change
- Willingness of the customer to take ownership of the project.

The ERP systems are modular in nature i.e their implementation should not be done in a single step. It can be divided into various stages like :

1. **Small size project :** It can be planned and delivered in 3-9 months.
2. **Medium size projects :** It can be done in 14 months and requires 150 consultants.
3. **Large size projects :** It requires years for their implementation.

11.8.1. Identification of Suitable Platform and Environment for ERP Implementation

For the implementation of an ERP system, suitable platform is required. The two major factors in ERP implementation are :

- Operating system
- Relational database

Operating system and database : Two main factors are taken into consideration :

- If the system is not created from the beginning, integration of the ERP system with the existing operating system and database is a better option. This helps in reducing the expenditure of the implementation process. But the major factor to be taken into consideration is the compatibility with the current system.
  The most popular database is Oracle, IBM, Microsoft and Sybase. In the OS market, popular operating systems are UNIX, IBM and Microsoft.
  But if the existing system is not compatible with the ERP package, alternatives should be used.
- It should be known whether the ERP systems are to be hosted in-house or hosted remotely. If the ERP system is hosted remotely, the cost of hardware, operating system and database is eliminated.

Networking Issues : The networking issues in an ERP implementation are :

- Network topology
- Network design
- Network printers

Setting up of a new or modifying an existing Local area network (LAN) and Wide area network (WAN) constitute an important part of the implementation process. The major functions are :

- Cabling to the planned workstation
- Providing sufficient bandwidth to the remote sites
- Providing fibre optics as the backbone

11.8.2. Phases of an ERP Implementation

There are various phases of an ERP implementation. Each phase has an entry criterion. This entry criteria needs to be fulfilled before the phase can start.

1. Phase 1 : Project Establishment :

The project establishment comprises of the following functions :

- Mobilization
- Project team creation
- Project plan detailing

For the creation of the project, several skills are required. For this purpose, different teams are created. This team consists of the members from the organization as well as the vendor.

- Project board
- Key user team
- Consultant's team
- IT personnel
- Project implementation committee

(*a*) Entry Criteria : These are :

- Vendor project team on site

- Organization's implementation team

(b) Deliverables : These are :

- Project administration plan
- Project standard document
- Project Implementation Committee
- Key User Team Formation
- Delivery of ERP application
- Installation of ERP application

2. Phase 2 : Procedure Development

In this phase, the key users and the team map the existing functional processes with the functions that are available in the ERP application. The activities which take place are :

- Process mapping
- Gap detection
- Organizational impact analysis

(a) Deliverables :

- Business model through flow chart
- Draft procedure manual
- Acceptance test plan and data
- Design specification document

3. Phase 3 : Application Software Customization

This phase includes the program development for all modules. Each program will be unit tested. The developed and unit tested application is then system tested. The system testing is based on the acceptance test plan and the test data that is provided by an organization.

(a) Entry Criteria :

- Acceptance test plan
- Acceptance test data
- Design specification document

(b) Deliverables :

- Customized unit and tested application software

4. Phase 4 : Acceptance Testing

Acceptance testing is the final quality control procedure to determine whether the software product is performing as per expectations.

(a) Entry Criteria :

- Availability of system tested application

(b) Deliverables :

- Updated procedure manual
- Tested application software

5. Phase 5 : End User Training

End users are trained in the specific functions they are required to work upon. The training is conducted by the vendor.

(a) Entry Criteria :

- Tested application software
- Updated procedure manual

(b) Deliverables :

- End users are trained as per the training plan

6. Phase 6 : Conference Room Trial

It is the model of the actual business environment. Activities which occur during this phase are :

- Identification of the representative data
- Entry of the representative data
- Creation of the business model
- Improvement of the business model
- Finalizing the procedure manual

(a) Entry criteria :

- Tested application software
- Updated procedure manual

(b) Deliverables :

- Final procedure manual
- Final business model

7. Phase 7 : Data Loading

Data loading is the migration phase where data from the existing manual is transferred to the system's database. It makes use of the interface programs that are available within the ERP package. Following activities takes place :

- Data conversion
- Data loading
- Establishment of access rights
- Establishment of passwords

(a) Entry criteria :

- Final procedure manual
- Final business model

(b) Deliverables :

- Data loading completed

8. Phase 8 : Live Trail

Once the final business model and the database is ready, the project can go live. The live trail phase allows the users to work upon the application software with real time data :

(a) Entry criteria :

- Data loading completed
- Trained end users

(b) Deliverables :

- Live trail completed

11.8.3. Data Migration

Data migration is one of the most important characteristic in determining the success of an ERP system. The steps involved in this approach are listed below :

- Identify the data to be migrated
- Determine the time taken for performing data migration

- Generate data templates
- Reserve the tools for data migration
- Decide on migration related setups
- Decide on data archiving

11.8.4. Customization vs Configuration

ERP vendors have tried to reduce the configuration by providing built in configuration tools. The key differences are :

Customization	Configuration
1. It is optional.	1. Some degree of configuration is required before the working of the software.
2. It is available to individual customers.	2. It is available to all the customers.
3. The effect of customization on the performance of the system is unpredictable.	3. The effect of configuration on the performance of the system is predictable.

11.9. ERP KEY VENDORS

The ERP key vendors are listed below :

(*a*) **SAP :** SAP is the largest ERP solution provider. SAP has a total of 12 million users and holds 30 % of the market share. It is available in all the major currencies and languages. It can also be hosted on almost every operating system and database. The SAP offering for smaller organizations is SAP Business One.

For example: "SAP business by design" is a solution for the organizations that lack IT resources.

(*b*) **Oracle :** Oracle is next to SAP. It offers multilanguage and multicurrency solution through its solution partners. Oracle is the first to provide Internet computing model for the development and deployment of its products.

(*c*) **Infor :** Infor is a recent ERP vendor. It focusses on building, buying and integrating the best solutions.

(*d*) **Microsoft Dynamics :** Microsoft Dynamics acquired a host of ERP products like Navision, Great Plain etc. It is dependent on its channel partners for sales, consulting and add-on development.

11.10. COMMERCIAL APPLICATIONS

The commercial applications of ERP are :

1. **Manufacturing :** It includes engineering, scheduling, work orders, capacity, quality control and cost management.
2. **Supply chain management :** It includes the supply chain planning, purchasing, inventory and order entry.
3. **Financials :** It includes cash management, accounts payable and account receivable.
4. **Project management:** It includes costing and billing, time and performance.
5. **Customer Relationship Management:** It includes the sales and marketing, call centre support and service provision.

11.11. ERP LIFECYCLE

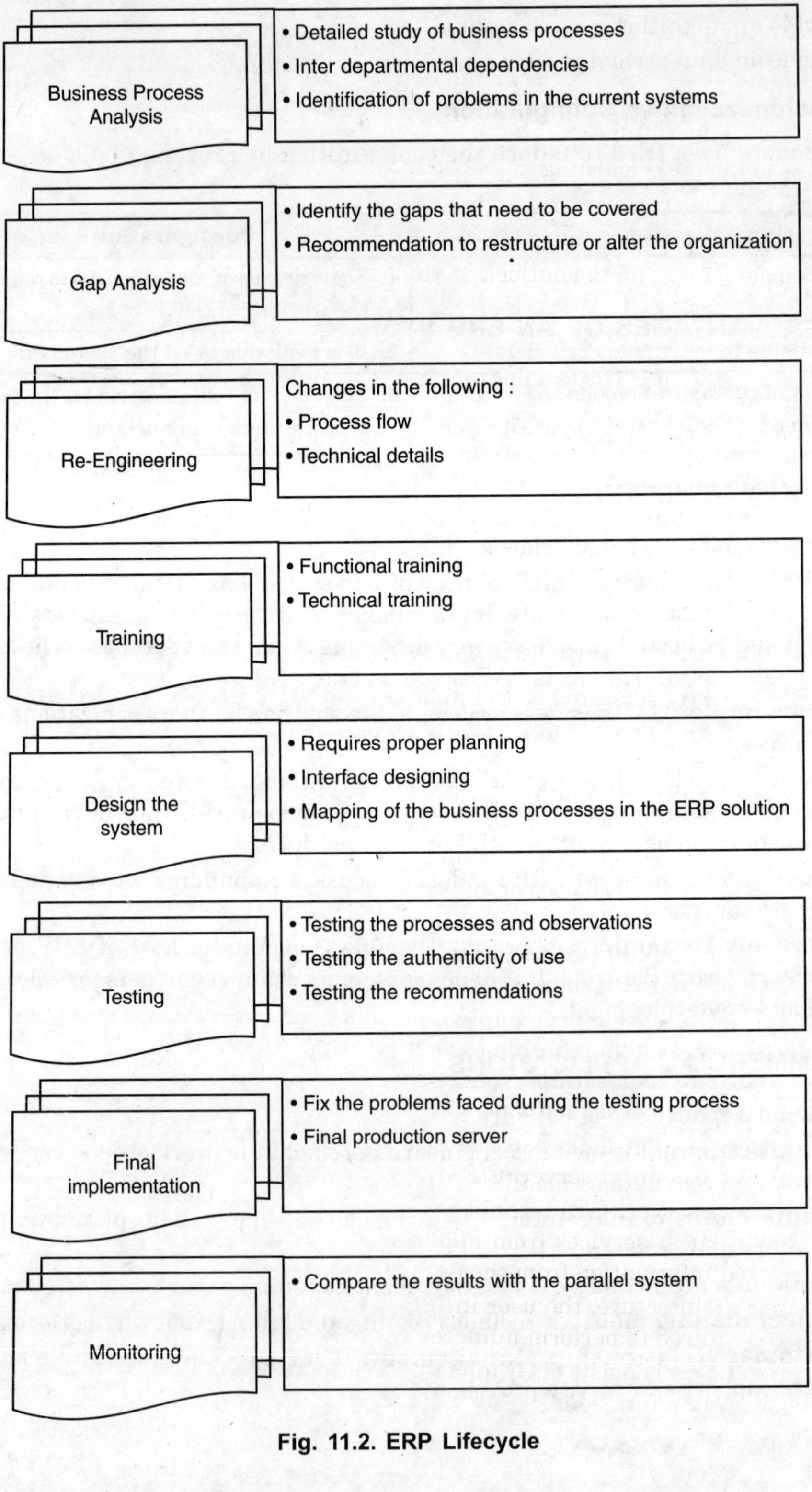

Fig. 11.2. ERP Lifecycle

11.12. ADVANTAGES OF AN ERP SYSTEM

The advantages of an ERP system are :

(*a*) It provides integration among different functional areas to ensure proper communication, productivity and efficiency.

(*b*) It helps to deliver the best product.

(*c*) Reduces delivery time

(*d*) Helps in business learning

(*e*) Efficient

(*f*) It helps the management to take important decisions with an in-depth study.

(*g*) It helps in changing the outlook of any business organization.

11.13. DISADVANTAGES OF AN ERP SYSTEM

The disadvantages of an ERP system are :

(*a*) It is expensive.

(*b*) Customization of ERP software is limited.

(*c*) It is rigid in nature.

(*d*) Resistance in sharing critical information between departments can result in reducing the effectiveness of the software.

Summary

- Enterprise resource planning is an integrated computer based application that is used to manage internal and external resources, financial resources, materials and human resources.
- Material requirements planning (MRP) system is the planning of the product or parts requirement according to the master production schedule.
- The main purpose of an ERP is to facilitate the flow of information between all business functions inside the boundaries of an organization.
- Integration of system, enable new business to grow, reduce expense on business processes, effective decision making, technological advancement, improved communication and easy management of the HR resources are the few reasons for implementing ERP.
- Customization, maintenance, implement-ability, real time changes, flexibility, user friendliness, cost, system requirements, after sales support and system, back-up system, reporting and analyzing features, vendor credentials, Internet integration, financing options and integration with other software's are the selection criteria of an ERP process.
- An Enterprise system makes use of two technologies namely thin client/server (C/S) technology and Client/fat server (C/FS) architecture.
- In a Thin client/server (C/S) technology, a number of client devices are operated by end users. They request services from application servers, which in turn get the requested service-related information from the database servers.
- In three tier architecture, the user interface runs on the client. Here, powerful PC's and servers are required to perform numerous operations.
- Presentation Layer consists of Graphical user interface (GUI), browser for data entry and accessing of system functions.

- Application Layer consists of Business rules, functions, logic and programs which act on data that is received or transferred from and to the database servers.
- Database Layer is the management of the organization's operational or transactional data including the metadata. It mostly employs industry standard RDBMS with structured query language (SQL) provisions.
- Transactional backbone, advanced application and management portal are the components of an ERP.
- Project establishment, procedure development, application software customization, acceptance testing, end user training, conference room trial, data loading and live trail are the phases of an ERP implementation.
- Business process analysis, GAP analysis, Re-engineering, training, design the system, testing, final implementation and monitoring are the phases of an ERP lifecycle.
- SAP, Oracle, Infor and Microsoft dynamics are the key vendors of ERP.
- Manufacturing, supply chain management, financials, project management and customer relationship management are the commercial applications of ERP.

Short Question Answers

1. Define Enterprise Resource Planning.

Ans. Enterprise resource planning is an integrated computer based application that is used to manage internal and external resources, financial resources, materials and human resources. **American Production and Inventory Control Society (2001)** has defined ERP systems as a: "Method for an effective planning and controlling of all the resources needed to take, make, ship and account for customer orders in a manufacturing, distribution or service company."

2. What are the features of an ERP?

Ans. The features of an ERP are :

- It is built on a centralized database.
- It makes use of a common computing platform.
- It can reside on a centralized server or can be distributed across a modular software or hardware unit that provides services and communicate on a LAN.
- Distributed : It allows business to assemble modules from different vendors without the need for fixing of multiple copies of complex and expensive computer systems.

3. What are the purposes of an ERP?

Ans. The main purposes of an ERP are :

- Improve productivity
- Enable easy communication with clients and partners
- Integration of systems
- Provide flexibility
- It helps in making quick and vital decision making.
- Improving the return on investment

4. What are the reasons for implementing ERP?

Ans. Integration of system, enable new business to grow, reduce expense on business processes, effective decision making, technological advancement, improved communication and easy management of the HR resources are the few reasons for implementing ERP.

5. What are the criterions of ERP selection process?

Ans. Customization, maintenance, implement-ability, real time changes, flexibility, user friendliness, cost, system requirements, after sales support and system, back-up system, reporting and analyzing features, vendor credentials, Internet integration, financing options and integration with other software's are the selection criteria of an ERP process.

6. Explain the ERP system architecture?

Ans. **Thin client/server (C/S) technology :** In a C/S system, a number of client devices are operated by the end users. They request services from application servers, which in turn get the requested service-related information from the database servers.

Three-tier architecture : In three tier architecture, the user interface runs on the client. Here, powerful PC's and servers are required to perform numerous operations.

7. What are the three layers of logic in three tier architecture?

Ans. The three layers of logic in three tier architecture are :

1. **Presentation Layer :** It consists of Graphical user interface (GUI), browser for data entry and accessing of system functions.
2. **Application Layer :** It consists of Business rules, functions, logic and programs which act on data that is received or transferred from and to the database servers.
3. **Database Layer :** It is the management of the organization's operational or transactional data including the metadata. It mostly employs industry standard RDBMS with structured query language (SQL) provisions.

8. What are the components of an ERP?

Ans. Transactional Backbone, advanced application and management portal are the components of an ERP.

9. What are the commercial applications of ERP?

Ans. Manufacturing, supply chain management, financials, project management and customer relationship management are the commercial applications of ERP.

10. What are the networking issues in ERP implementation process?

Ans. The networking issues in an ERP implementation are :

- Network topology
- Network design
- Network printers

11. State the ERP vendors.

Ans. Oracle, Infor, SAP and Microsoft dynamics are the ERP vendors.

Review Questions

1. Explain ERP and its evolution.
2. Explain ERP system architecture in detail.
3. Explain the implementation of ERP in detail.
4. Explain the lifecycle of ERP.
5. Explain the advantages and disadvantages of ERP.

❑❑❑

Chapter 12

Supply Chain Management

"The opportunity to engage people who are very far away from each other and yet might have interesting information is better here than has ever been possible."

Inside this Chapter

12.1. SUPPLY CHAIN MANAGEMENT OVERVIEW

Supply chain management are a set of approaches that are used to efficiently integrate the suppliers, manufacturers, warehouses and distribution centres. It is also the management of a network of interconnected businesses that are involved in the delivery of product and service packages that is required by the end customers.

The approaches are performed so that the product is produced and distributed in :

- Right quantity
- Right location
- Right time

Supply chain management is also known as logistics network.

Supply chain management is the designing, planning, controlling, monitoring of the supply chain activities with an objective of creating net value, measuring performance globally and synchronizing supply with demand.

12.2. EVOLUTION OF SUPPLY CHAIN MANAGEMENT

The evolution of supply chain management consists of six major movements. These are :

I. Creation Era : The word supply chain management was invented by the U.S. industry consultant in 1980's. The characteristics of the creation era are :

- Large scale changes
- Re-engineering
- Cost reduction programs

II. Integration Era : Integration era came into existence with the development of electronic data interchange in the 1960's. In the late 1990's, Enterprise Resource Planning was developed. The integration era was developed in the 21st century with the expansion of Internet based collaborative systems. The characteristics of the integration era are :

- Cost reduction
- Increased value-adding

Classification of Supply Chain : The supply chain can be classified as :

- **Stage 1 Network :** In stage 1 network, various systems like make, storage, distribution and material control are independent of each other.
- **Stage 2 Network :** In stage 2 network, various systems like make, storage, distribution and material control are dependent of each other. These are integrated under one plan and are ERP enabled.
- **Stage 3 Network :** In stage 3 network, vertical integration is achieved with the suppliers in upstream direction and customers in downstream direction. **For example**: Supply chain – Tesco.

> **DO YOU KNOW?**
>
> "The information about a package is as important as the delivery of the package itself."
>
> –Frederick W.Smith (1979)

III. Globalization Era : The characteristic of the globalization era is the attention provided to the global systems of supplier relationships and the expansion of supply chains over national boundaries. The characteristics of the globalization era are :

- Competitive advantage
- Value adding
- Reducing costs through global sourcing

IV. Specialization Era (Phase I) : In the 1990's, specialization era came into existence. Industries adopted the specialization model. The specialization model creates manufacturing and distribution networks. These networks are composed of multiple individual supply chains that are specific to :

- Products
- Suppliers
- Customers

These individuals work together to design, manufacture, market, sell and service a product.

V. Specialization Era (Phase II) : The supply chain specialization allows companies to improve their overall competencies. It helps to increase the performance and efficiency of the supply chain.

VI. Supply Chain Management 2.0 : Supply chain management 2.0 describes the following :

- Changes within the supply chain
- Evolution of the processes

- Evolution of the methods
- Evolution of the tools

The solutions are delivered in many ways such as :

- **No-Touch** through business process outsourcing.
- **Mid-touch** management services and software as a service (SaaS).
- **High Touch** through traditional software deployment model.

12.3. INPUTS OF SUPPLY CHAIN MANAGEMENT

There are three inputs of supply chain management. These are :

- Raw material
- Working in inventories
- Finished products

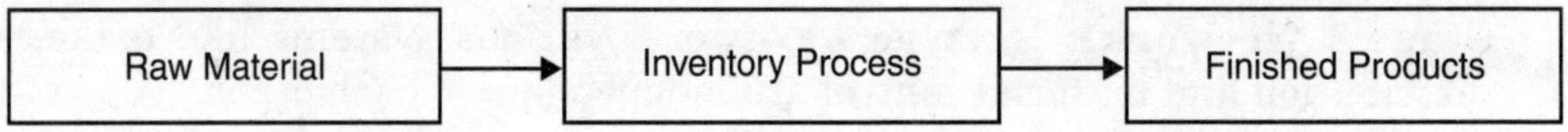

Fig. 12.1. Inputs of Supply Chain Management

12.4. SUPPLY CHAIN STRUCTURE

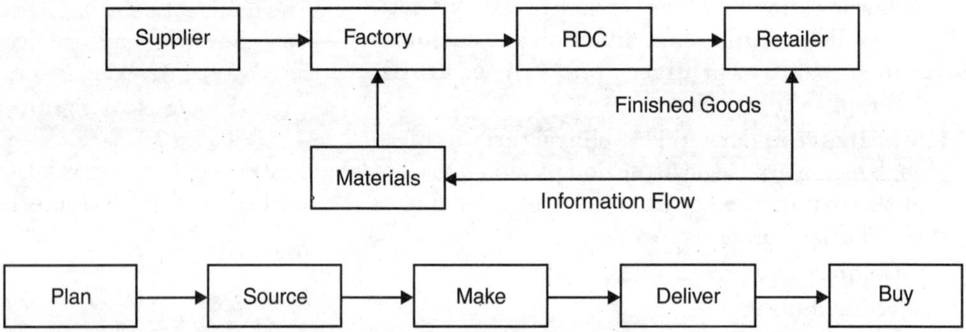

Fig. 12.2. Supply Chain Structure

12.5. SUPPLY CHAIN ELEMENTS

There are three supply chain elements. These are strategic, tactical and operational elements.

1. **Strategic Elements :** The strategic elements are the following :
   - **Strategic network optimization :** It includes the number, location and size of the distribution centres.
   - **Strategic partnerships :** This is the partnership with the suppliers, distributors and customers. It provides a communication channel for sharing critical information.
   - **Product lifecycle management :** It is the integration of the new products with the existing ones into the supply chain activities.
   - Supply chain design
   - Long term planning
   - Resource attainment
2. **Tactical Elements :** The tactical elements are the following :
   - **Production decisions :** These include contracting and scheduling.

- Distribution planning
- Production planning
- Medium term planning
- Resource allocation
- Purchasing decisions
- **Inventory decisions :** These include quantity, location and quality of the inventory.
- **Transportation strategy :** These include routes and contracting.
- Customer demands and habits.

3. **Operational Elements :** The operational elements are the following :
   - Shipment scheduling
   - Short term planning
   - Resource scheduling
   - Production scheduling
   - Demand planning and forecasting
   - **Inbound operations :** It includes the transportation from suppliers.
   - **Outbound operations :** It includes warehousing and transportation to the customers.
   - **Production operations :** It includes the consumption of materials and flow of finished goods.
4. **Management Elements :** The management elements are the following :
   - Planning and control
   - Work structure
   - Organization structure
   - Risk and reward structure
   - Culture and attitude
   - Management methods
   - Power and leadership structure

12.6. SUPPLY CHAIN MANAGEMENT STRATEGIES

The supply chain management strategies are of two types. These are operation and integration strategies.

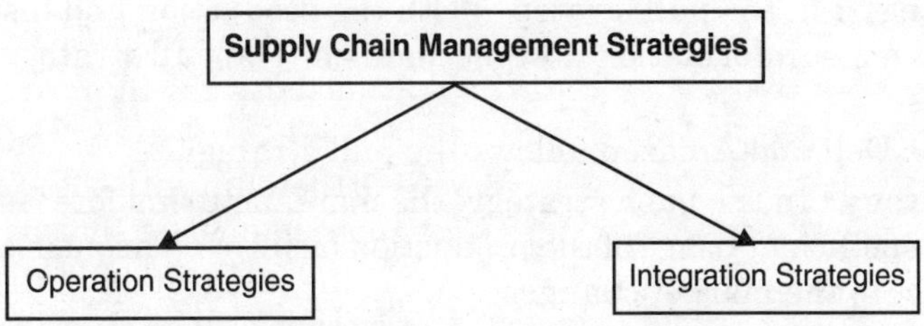

Fig. 12.3. Supply Chain Management Strategies

12.6.1. Operation Strategies

The operation strategies are of three types. These are :

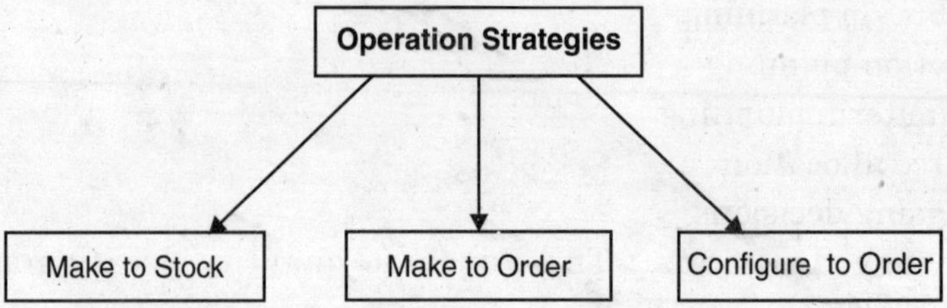

Fig. 12.4. Operation Strategies

I. Make to Stock : The stock must be chosen when the :

- Products are standardized and
- Demand is predictable.

The benefits of make to stock are that the manufacturing costs are lower and the demands of the customers are fulfilled quickly.

II. Make to Order : The order must be chosen when the :

- Products are customized.

The benefits of make to order are that it improves the security levels and reduces the inventory cost.

III. Configure to Order : The order must be chosen when there are :

- Various variations on the finished products.
- Demand is infrequent.

The benefits of configure to order are low inventory costs and simple planning.

12.6.2. Integration Strategies

The supply chain integration strategies are of three types. These are :

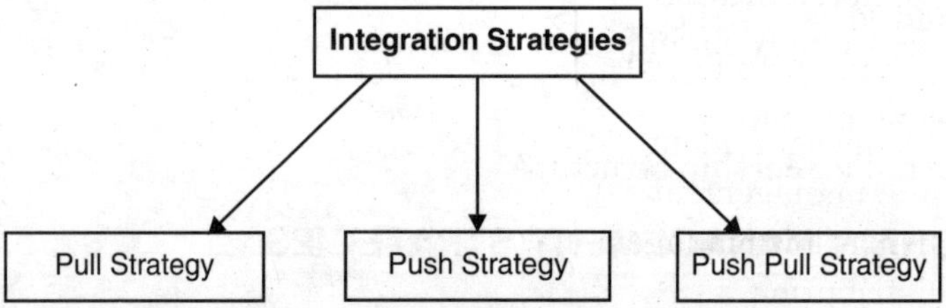

Fig. 12.5. Integration Strategies

I. Pull Strategy : In the pull strategy, both the production and distribution is demand driven. It is a fast information flow mechanism. The pull strategy makes efficient use of resources.

Examples : Dell and Amazon follows the pull strategy.

II. Push Strategy : In the push strategy, the manufacturing forecasts are of long range. There is an inefficient usage of the production facilities. The push strategy takes longer time to react to the market changes.

Example : Auto Industry.

III. Push | Pull Strategy : The push pull strategy is the hybrid of push and pull strategy. This strategy overcomes the disadvantage of both the push and pull strategy. The early stage of product assembly is done with the help of **push** strategy. The final staging is done with the help of **pull** strategy. The advantage of using this strategy is that it helps to reduce uncertainty. The final product assembly done is on the demand of the customers.

Example : Cisco Systems

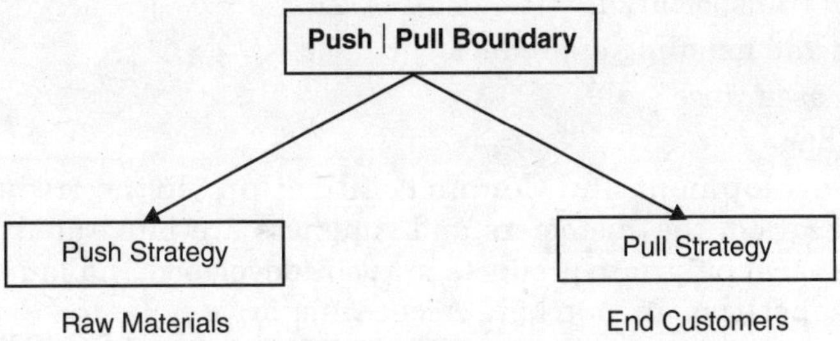

Fig. 12.6. Supply Chain Timeline

12.7. BUSINESS PROCESS INTEGRATION

Supply chain business process integration is the collaboration between buyers and suppliers. According to **Lambert and Cooper**, integrated supply chain requires a continuous information flow. The key supply chain processes stated by Lambert are :

- Customer relationship management
- Customer service management
- Demand management style
- Order fulfilment
- Manufacturing flow management
- Supplier relationship management
- Return management
- Product development
- Product commercialization

Other key critical supply chain business processes started by Lambert are :

- Procurement
- Performance measurement
- Warehousing management
- Manufacturing flow management
- Physical distribution

I. **Customer Service Management Process :** Customer service management process is concerned with the relationship between the organization and its customers. It provides the customer with information on scheduling and product availability through the company's production and distribution companies. Following steps are taken by an organization to build customer relationships :
   - Determine mutually satisfying goals for the organization and customers
   - Establish and maintain customer relationship
   - Produce positive feelings in the organization and the customers

II. **Procurement Process :** In the procurement process, strategic plans are formed to support the manufacturing flow management process and the development of new products. It helps in the reduction of time that is required for the design cycle and product development. The activities that are related with the outside suppliers are :
   - Resource planning
   - Supply sourcing

- Order placement
- Inbound transportation
- Storage and handling of products
- Quality assurance
- Scheduling

III. Product Development and Commercialization : In the product development and commercialization, the customers and suppliers are integrated. As the product life cycle shortens, appropriate products must be developed and launched in less time to remain competitive in market. According to Lambert and Cooper, managers of the product development and commercialization must take the following actions into consideration :

- Coordinate with the customer relationship management to identify the customer needs.
- Select the materials and suppliers in combination with the procurement.
- Develop production technology. Integrate the technology with the supply chain.

DO YOU KNOW?

"Part of what you need to do in the supply chain is to help your company anticipate events, and understand the environment you operate in – physical, political, ecnomic – around the globe."

IV. Manufacturing Flow Management Process : The manufacturing flow management process produces and supplies products to the distribution channels based on the previous forecasts. The processes must be flexible to respond to the changes occurring in the market. The advantages of these processes are :

- Improved responsiveness
- Efficiency in meeting the demands of the customer

V. Physical Distribution : In physical distribution, the customer is the final destination of a marketing channel. The main concern in the physical distribution is the movement of the finished products and services to the customers. The time and space of the customer service are an important part of the marketing.

VI. Outsourcing : Outsourcing is the procurement of the following :

- Materials
- Components
- Services

The major benefit of using this technique is that the company will now focus on those activities that have an advantage. It is beneficial in logistics where there is a provision of transport, warehousing and inventory control. The managing and controlling of the network partners and suppliers requires both central and local movement.

VII. Performance Measurement : A. T Kearney Consultants noted that the firms that participate in performance measurement noted improvements in the overall productivity of the organization. The internal measures are collected and analysed by the firm. These are :

- Cost
- Customer service
- Productivity measures
- Asset management
- Quality

The external measures are examined by the customer. These measures are :

- Best practice benchmarking
- Customer perception measurement

VIII. Warehousing Management : Warehousing management helps in performing the following operations :

- Reducing the company costs and expenses
- Reducing manpower cost
- Dispatching products on time
- Loading facilities with proper area
- Unloading facilities with proper area

12.8. KEY ISSUES IN SUPPLY CHAIN MANAGEMENT

There are five main key issues of a supply chain management. These are :

(*a*) **Forecasts are never right :** It is doubtful that the actual demand will exactly equal the forecasted demand of a product.

(*b*) **Longer the forecast horizon, the worse is the forecast :** A forecast that is done for a year from now will never be as accurate as a forecast that is done for three months from now.

(*c*) Aggregate forecasts are more accurate.

(*d*) **Network Planning :** The network planning includes the following :

- Warehouse location
- Warehouse capacity
- Plant location
- Production level
- Transport facility

(*e*) **Supply Contracts :** The supply contract includes the impact made by providing volume discount and revenue sharing.

12.9. METHODS FOR IMPROVING FORECASTS

For achieving accurate forecast, there are four methods for improving the forecasts. These methods are :

(*a*) **Judgement Methods :** This method consists of the following operations :

- Internal experts
- External experts
- Domain experts

(*b*) **Market Research Analysis :** This method consists of the following operations :

- Market testing
- Market survey

(*c*) **Time Series Methods :** This method consists of the following operations :

- Trend analysis
- Seasonality analysis

(*d*) **Casual Analysis :** This method consists of the following operations :

- Economic data
- It relies on the data that is not predicted.

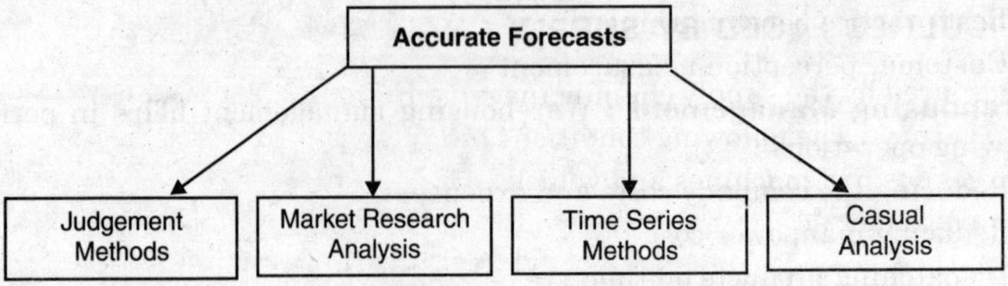

Fig. 12.7. Forecasting Methods

12.10. SUPPLY CHAIN MANAGEMENT THEORIES

Currently, there is no theoretical explanation available for the existence and boundaries of supply chain management. Few authors like Halldorsson, Ketchen and Hult, Lavassani tried to provide theoretical foundations for the different areas of the supply chain management.

Some of the theories are :

(*a*) Resource based view
(*b*) Transaction cost analysis
(*c*) Knowledge based view
(*d*) Strategic choice theory
(*e*) Agency theory
(*f*) Institutional theory
(*g*) Systems theory
(*h*) Network perspective
(*i*) Materials logistics management
(*j*) Just in time
(*k*) Material requirement planning
(*l*) Theory of constraints
(*m*) Total quality management
(*n*) Time based competition
(*o*) Quick response manufacturing
(*p*) Available to promise

12.11. TAX EFFICIENT SUPPLY CHAIN MANAGEMENT

Tax Efficient Supply Chain Management is basically a business model that considers the effect of tax in the design and implementation of the supply chain management. The business organizations that are in different countries should pay different tax rates in different countries. This is due to the effect of globalization. Due to this difference, global business organizations have the opportunity to calculate and optimize the supply chain. It is thus used as a method of gaining more profit for company which owns global supply chain.

12.12. BENEFITS OF SUPPLY CHAIN MANAGEMENT

The benefits of the supply chain management are :

- Delivery performance
- Inventory reduction
- Lower supply chain costs
- Service level requirements are satisfied

12.13. DIFFICULTIES FACED BY SUPPLY CHAIN MANAGEMENT

The difficulties faced by the supply chain management are :

1. **Uncertainty :** The following conditions are there :
   - Breakdown of machines and vehicles
   - Travel time
   - Weather
   - Labour conditions
2. **Complexity :** The following conditions are there :
   - Minimize internal costs
   - Minimize uncertainty
3. **Distribution Network Configuration :** The following conditions are there :
   - Number of suppliers
   - Location of suppliers
   - Network of suppliers
   - Production facilities
   - Distribution centres
   - Warehouses
   - Customers
4. **Distribution Strategy :** The following conditions are there :
   - Questions of operating control *i.e.,* centralized, decentralized or shared.
   - Delivery scheme i.e. direct shipment, pool point shipping, direct store delivery, closed loop shipping.
   - Mode of transportation i.e. motor carrier, truckload, parcel, railroad.
   - Replenishment strategy i.e. pull, push or hybrid.
   - Transportation control i.e. private, common and contract carrier
5. **Information :** The following conditions are there :
   - Demand signals
   - Forecasts
   - Inventory
   - Transportation
   - Collaboration
6. **Inventory Management :** The following conditions are there :
   - Quantity of inventory
   - Location of inventory
   - Work-in-process
   - Finished goods
7. **Cash-Flow :**The following conditions are there :
   - Arranging the payment terms
   - Methodologies for exchanging funds across entities within the supply chain

12.14. APPLICATIONS OF SUPPLY CHAIN MANAGEMENT

The applications of the supply chain management are :

- Globalization

- Increased cross border sourcing
- Collaboration for parts of value chain with low-cost providers
- Shared service centres for logistical and administrative functions
- Increasingly global operations which requires increasingly global coordination and planning.

Summary

- Supply chain management are a set of approaches that are used to efficiently integrate the suppliers, manufacturers, warehouses and distribution centres.
- Supply chain management is also known as logistics network.
- Supply chain management is the designing, planning, controlling, monitoring of the supply chain activities with an objective of creating net value, measuring performance globally and synchronizing supply with demand.
- Integration era came into existence with the development of electronic data interchange in the 1960's.
- The characteristic of the globalization era is the attention provided to the global systems of supplier relationships and the expansion of supply chains over national boundaries.
- The supply chain specialization allows companies to improve their overall competencies. It helps to increase the performance and efficiency of the supply chain.
- The three inputs of supply chain management are raw material, working in inventories and finished products.
- Strategic, tactical, operational and management are the supply chain elements.
- Operation and integration are the two types of supply chain management strategies.
- Make to stock, make to order and configure to order are the three types of operation strategies.
- Push, pull and push | pull are the three types of integration strategies.
- In the pull strategy, both the production and distribution is demand driven.
- In the push strategy, the manufacturing forecasts are of long range.
- The push | pull strategy is the hybrid of push and pull strategy. This strategy overcomes the disadvantage of both the push and pull strategy.
- Supply chain business process integration is the collaboration between buyers and suppliers.
- Customer service management process is concerned with the relationship between the organization and its customers.
- In the procurement process, strategic plans are formed to support the manufacturing flow management process and the development of new products.
- In the product development and commercialization, the customers and suppliers are integrated.
- The manufacturing flow management process produces and supplies products to the distribution channels based on the previous forecasts.
- In physical distribution, the customer is the final destination of a marketing channel.
- Judgement methods, market research analysis, time series methods and casual analysis are the four methods for improving the forecast of supply chain management.
- Tax Efficient Supply Chain Management is basically a business model that considers the effect of tax in the design and implementation of the supply chain management.

Short Question Answers

1. Define supply chain management.

Ans. Supply chain management are a set of approaches that are used to efficiently integrate the suppliers, manufacturers, warehouses and distribution centres. It is also the management of a network of interconnected businesses that are involved in the delivery of product and service packages that is required by the end customers.

2. Explain the evolution of supply chain management in brief.

Ans. The evolution of supply chain management consists of six major movements. These are creation era, integration era, globalization era, specialization era (phase-1), specialization era (phase-2) and supply chain management 2.0.

3. Explain the creation era.

Ans. The word supply chain management was invented by the U.S. industry consultant in 1980's. The characteristics of the creation era are large scale changes, re-engineering and cost reduction programs.

4. Explain the integration era.

Ans. Integration era came into existence with the development of electronic data interchange in the 1960's. The characteristics of the integration era are cost reduction and increased value-adding.

5. Classify the supply chain.

Ans. The supply chain can be classified as :

- **Stage 1 Network :** In stage 1 network, various systems like make, storage, distribution and material control are independent of each other.
- **Stage 2 Network :** In stage 2 network, various systems like make, storage, distribution and material control are dependent of each other. These are integrated under one plan and are ERP enabled.
- **Stage 3 Network :** In stage 3 network, vertical integration is achieved with the suppliers in upstream direction and with customers in downstream direction.

6. Explain the globalization era.

Ans. The characteristic of the globalization era is the attention provided to the global systems of supplier relationships and the expansion of supply chains over national boundaries. The characteristics of the globalization era are competitive advantage, value adding and reducing costs through global sourcing.

7. Explain the specialization era (Phase-I).

Ans. In the 1990's, specialization era came into existence. Industries adopted the specialization model. The specialization model creates manufacturing and distribution networks. These networks are composed of multiple individual supply chains that are specific to the products, suppliers and customers. These individuals work together to design, manufacture, market, sell and service a product.

8. Explain the specialization era (Phase-II).

Ans. In the specialization era phase- II, the supply chain specialization allows companies to improve their overall competencies. It helps to increase the performance and efficiency of the supply chain.

9. Explain supply chain management 2.0.

Ans. Supply chain management 2.0 describes the following :

- Changes within the supply chain
- Evolution of the processes
- Evolution of the methods
- Evolution of the tools

10. What are the inputs of Supply Chain Management?

Ans. Raw material, working in inventories and finished products are the three inputs of Supply Chain Management.

11. What are the four supply chain elements?

Ans. Strategic, operational, tactical and management are the four supply chain elements.

12. What are the strategic elements?

Ans. Strategic network optimization, strategic partnerships, product lifecycle management, supply chain design, long term planning and resource attainment are the strategic elements.

13. What are the tactical elements?

Ans. Production decisions, distribution planning, production planning, medium term planning, resource allocation, purchasing decisions, inventory decisions and transportation strategy, customer demands and habits are the tactical elements.

14. What are the operational elements?

Ans. Shipment scheduling, short term planning, resource scheduling, production scheduling, demand planning and forecasting, inbound operations, outbound operations, production operations are the operational elements.

15. What are the management elements?

Ans. Planning and control, work structure, organization structure risk and reward structure are the management elements.

16. What are the two supply chain management strategies?

Ans. Operation and integration are the two supply chain management strategies.

17. What are the three operation strategies?

Ans. Make to stock, make to order and configure to order are the three types of operation strategies.

18. What are the three integration strategies?

Ans. Push, pull and push | pull are the three types of integration strategies.

19. Explain the pull strategy.

Ans. In the pull strategy, both the production and distribution is demand driven. It is a fast information flow mechanism. The pull strategy makes efficient use of resources. Dell and Amazon follows the pull strategy.

20. Explain the push strategy.

Ans. In the push strategy, the manufacturing forecasts are of long range. There is an inefficient usage of the production facilities. The push strategy takes longer time to react to the market changes. Auto Industry follows the push strategy.

21. Explain the push/pull strategy.

Ans. The push/pull strategy is the hybrid of push and pull strategy. This strategy overcomes the disadvantage of both the push and pull strategy. The early stage of product assembly is done with the help of push strategy. The final stage is done with the help of pull strategy. The advantage of using this strategy is that it helps to reduce the uncertainty. The final product assembly done is done on the demand of the customer.

22. Explain the customer service management process.

Ans. Customer service management process is concerned with the relationship between the organization and its customers. It provides the customer with information on scheduling and the product availability through the company's production and distribution companies.

23. Explain the procurement process.

Ans. In the procurement process, strategic plans are formed to support the manufacturing flow management process and the development of new products. It helps in the reduction of time that is required for the design cycle and product development.

24. Explain the product development and commercialization.

Ans. In the product development and commercialization, the customers and suppliers are integrated. As the product life cycle shortens, appropriate products must be developed and launched in less time to remain competitive in market.

25. Explain the manufacturing flow management process.

Ans. The manufacturing flow management process produces and supplies products to the distribution channels based on the previous forecasts. The processes must be flexible to respond to the changes occurring in the market.

26. What is physical distribution?

Ans. In physical distribution, the customer is the final destination of a marketing channel. The main concern in the physical distribution is the movement of the finished products and services to the customers. The time and space of the customer service are an important part of the marketing.

27. What are the key issues in supply chain management process?

Ans. There are five main key issues of a supply chain management. These are :

(*a*) Forecasts are never right

(*b*) Longer the forecast horizon, the worse is the forecast

(*c*) Aggregate forecasts are more accurate

(*d*) Network Planning

(*e*) Supply Contracts

28. What are the methods of improving forecast in supply chain management?

Ans. Judgement methods, market research analysis, time series methods and casual analysis are the four methods for improving the forecast of supply chain management.

29. Explain the judgement method.

Ans. Internal experts, external experts and domain experts are the three judgement methods.

30. Explain the market research analysis.

Ans. Market testing and market survey are the two market research analysis methods.

31. Explain the time series methods:

Ans. This method consists of the trend analysis and seasonality analysis methods.

32. What is tax efficient supply chain management?

Ans. Tax Efficient Supply Chain Management is basically a business model that considers the effect of tax in the design and implementation of the supply chain management. The business organizations that are in different countries should pay different tax rates in different countries. It is thus used as a method of gaining more profit for company which owns global supply chain.

Review Questions

1. Explain supply chain management in detail.
2. What are the supply chain management theories?
3. What are the advantages and disadvantages of supply chain management?
4. What are the applications of supply chain management?
5. Explain business process integration.
6. Explain the evolution of supply chain management.

□□□

Chapter 13

Customer Relationship Management

"The easiest kind of relationship is with ten thousand people, the hardest is with one."

Inside this Chapter

- Customer Relationship Management Overview
- Types of Customer Relationship Management
- Open Source Customer Relationship Management
- Customer Relationship Management Goal
- Customer Relationship Management vs. Traditional Marketing Techniques
- Customer Relationship Management Areas
- Stages in Customer Relationship Management
- Determinants of Customer Relationship Management
- Models of Customer Relationship Management
- Functions of Customer Relationship Management
- Variations of Customer Relationship Management
- Strategy of Customer Relationship Management
- Implementation issues of Customer Relationship Management
- Help Menus
- Adoption Rates of Customer Relationship Management
- Benefits of Customer Relationship Management
- Challenges faced by Customer Relationship Management
- Implementation of Customer Relationship Management
- Examples of Customer Relationship Management
- Technical Functionalities of Customer Relationship Management

13.1. CUSTOMER RELATIONSHIP MANAGEMENT OVERVIEW

Customer relationship management is the development and maintenance of mutually beneficial long term relationships with strategically significant customers. It is an IT improved value process which :

- Identifies
- Develops
- Integrates and
- Focusses on the various competencies of the firm to the "voice of customer" in order to develop a long term customer value.

Customer Relationship Management is also known as relationship marketing or customer management. Customer relationship management captures, analyses and distributes all relevant data from the customer to everyone in an organization. This distribution of information helps the organization in meeting the customer, product and service needs.

13.2. TYPES OF CUSTOMER RELATIONSHIP MANAGEMENT

There are three types of customer relationship management. These are :

1. Operational CRM : The operational CRM helps in easy and effective communication with the customers.

In this type of CRM, various tools are used known as **contact management tools**. The major benefits of these tools are:

- Reduces the cost
- Increases efficiency
- Makes use of media based communication channels
- Provides consistent interface across all the communication channels.

Here relevant customer data is collected and displayed at appropriate customer touch points. This is known as **customer master data**. There is another set of data which consists of information like hobbies, interests, details about children etc. This set of data is known as **soft customer data**. The example of operational CRM is banks.

Channel management tools aims to understand how the customer interacts with the company. The aim of the channel management tool is to deliver the products and services across multiple channels in an effective, efficient and consistent manner. The various processes that are managed by the channel management tools are :

- Campaign management
- Sales management
- Service management
- Complaint management

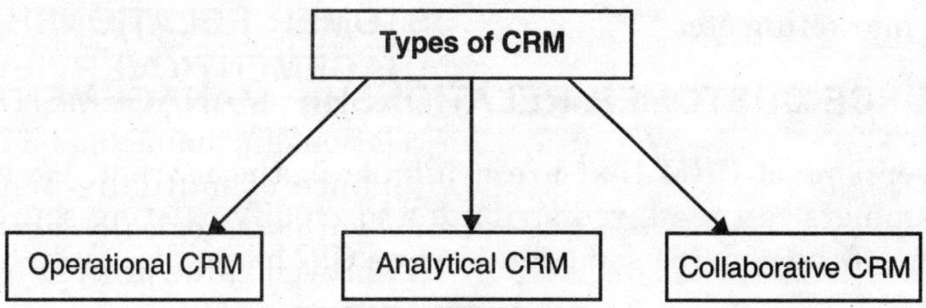

Fig. 13.1. Types of CRM

2. Analytical CRM : Analytical CRM is composed of the following :

- Pattern discovery component
- Product analysis component
- Customer analysis component
- Multitude component
- Sorting and customer fractionation component
- Customer value evaluation component

The analytical solutions are an integrated view of :

- Customers across all the channels
- Customers across all the applications
- Campaign performance analysis
- Cross selling
- Customer profit analysis

The analytical CRM solution helps in answering the following questions :

- Who are the best customers of an organization?
- Which customers an organization can loose?
- How to bind the customers with the organization?
- How to attract new customers?

- How to improve the profitability of customers?

The examples of analytical CRMs are :

- Data warehousing
- Online analytical processing(OLAP)
- Data mining systems

3. Collaborative CRM : The collaborative CRM allows the company to synchronize and manage efficient and productive interaction across all the communication channels with :

- Customers
- Prospects
- Partners
- Internal associates

In collaborative CRM, the viewpoint of the customer is taken into account at every transaction level. This helps in providing better services to the customer. Collaborative CRM helps in :

- Reducing web service costs with the help of web collaboration
- Improve the quality of customer service
- Increase customer loyalty

The example of collaborative CRM is the various departments of the company like sales, technical support, marketing etc.

13.3. OPEN SOURCE CUSTOMER RELATIONSHIP MANAGEMENT

There are many versions of CRM that are available in the market. With the help of open source, the programmers can read, redistribute and modify existing source codes to form a new software. Some of the widely used open sources CRMs are discussed below :

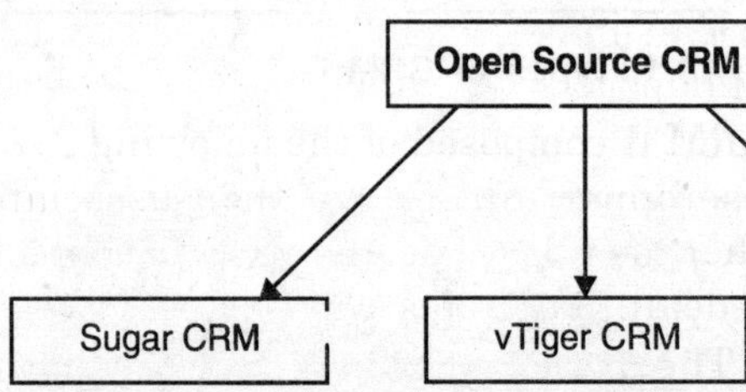

Fig. 13.2. Open Source CRM

1. **Sugar CRM :** Sugar CRM is a commercial open source CRM that is suitable for companies of all sizes. The version 3.0 is known as sugar suite. The various functions of this version of sugar CRM are :
   - **Sales Forecasting :** With the help of sales forecasting, the sales manager can calculate potential sales and the associated revenues. It provides a better control over performance, commitment, achieving goals and planning resources.
   - **Campaign and Project Management :** The campaign management tool allows users to perform the following functions :
     – Organize the marketing activities
     – Manage and track marketing efforts on leads, contacts and opportunities.

     The project management tool enables the users to :
     – Define steps
     – Establish timelines

– Assign work to members on any project that is useful to the end users.

- **Email Marketing :** The marketing team performs the following operations :
  – Mail templates
  – Listing of receivers
  – Sending personalized email templates as per the schedule.
- **Real Time Tracking :** The activities for real time tracking includes :
  – Calendaring
  – Call scheduling
  – Planning meetings
  – Note tracking
- **Document Management :** The documenting management includes capabilities like :
  – Managing sales documentation
  – Customer reports
- **Employee Directory :** The customers can be searched on the basis of :
  – First name
  – Last name
  – Roles
  – Department
  – Contact information

2. **VTiger CRM :** Vtiger CRM is the most suitable CRM for small and medium business organizations. It was originally built from the source code of Sugar CRM. This CRM is based on open source technologies like Linux, Apache, MySQL and PHP. The features are :
   - **Lead Management :** The lead management creates leads and then converts them to suitable opportunities. It imports important leads from Web, seminars and mails.
   - **Opportunity and Contact Management :** In this, the user can track all the sales opportunities. These opportunities are then associated with accounts, activities, contacts and other modules. The hierarchy of contacts is maintained so that the customers can be dealt in a proper way.
   - **Account Management :** The user can perform the following operations :
     – Import accounts
     – Export accounts
     – Specify parent-child relationship between accounts
     – Track all the accounts
     – Attach customer specific documents for quick reference
   - **Activity and Inventory Management :** With the help of activity and inventory management, the user can perform the following functions :
     – Add important customer related emails
     – Store all the details of customer meetings and calls
   - **Security Management :** The security management feature helps the organization in defining the role of team members and giving them access as per their role.
3. **Daffodil CRM :** Daffodil CRM is a commercial CRM. It helps to integrate all the aspects of customer life cycle. It helps in automating everyday tasks. The features are:
   - **Printable reports and Export facility :** The daffodil CRM has facilities of :

– Export to excel
– Printable report facilities

The view facility helps to generate different views of data. The users can evaluate the performance of their overall business, sales team and products over a period of time.

- **Automatic Mail Attachment :** Mails can be attached automatically to the daffodil CRM. Manual selection and sending is not required.
- **Campaign Facility :** It enables mass mailings to all or selected list of customers. Email templates may or may not be used for this purpose.
- **Lead Management :** Lead management helps the users in managing and analysing all leads within a single click. A single click provides :
  – All information related to the lead
  – Latest five actions and interactions for follow-up
  – Analysis made by the lead

13.4. CUSTOMER RELATIONSHIP MANAGEMENT GOAL

The goals of the customer relationship management are :

- It helps to find, attract and win new clients for the company.
- It helps to hold those clients that are already associated with the organization.
- It helps the organization in understanding each customer's value to the company in a better way.
- It must aim to reduce the cost of marketing and client service.
- It improves the efficiency and effectiveness of communication.

13.5. CUSTOMER RELATIONSHIP MANAGEMENT VS. TRADITIONAL MARKETING TECHNIQUES

There is a huge difference between the traditional techniques that were employed by the organization and the new technique i.e. customer relationship management. Some of them are :

- Traditional marketing techniques focussed on elements like :
  – Price
  – Product
  – Promotion
  – Place
- Traditional marketing techniques are functionally based. They neglect the customer after the sales process is over and fails to meet the customer needs.
- Customer Relationship Management takes care of the customer needs and is one of the most useful tools in the company's success.

> **DO YOU KNOW?** ✍
>
> "The first rule of any technology used in a business is that automation applied to an efficient operation will magnify the efficiency. The second is that automation applied to an inefficient operation will magnify the inefficiency."
>
> – Bill Gates

13.6. CUSTOMER RELATIONSHIP MANAGEMENT AREAS

There are three areas of the customer relationship management. These are :

1. **Sales :** The sales force predicts and proposes real time analysis of the information that is available and then distributes it to the company. It makes use of the software to

reorganize all the phases of the sales process. Thus it minimizes the time that is required by the sales representatives to spend on each phase. The main advantage of this technique is that is allows an organization to use fewer sales representative to manage their clients. In the sales area, there is a sales force automation. The core of the sales force automation is the contact management system. This system is used for tracking and recording each stage in the sales process.

2. **Marketing :** The marketing area concentrates on personalizing the customer prefernces and offer them satisfying experiences. It helps the enterprise to identify and target the possible clients and helps the sales team. The main advantages in this area are :
   - Email search
   - Social media
   - Telephone
   - Direct mails
3. **Service :** The service area is associated with the following :
   - Call centers
   - Email
   - Other communication medias

   The call center software helps to direct the customer to the agent that can help them with their current problems. Organizations are considering and using this technology to attract and retain their customers. It helps them to :
   - Improve their clients experience
   - Increases efficiency
   - Minimize costs

   Customer relationship management helps in building value driven long term relationships.

> **DO YOU KNOW?**
>
> "A business absolutely devoted to Customer Service Excellence will have only one worry about profits. They will be embarrassingly large."
>
> –Sir Henry Ford

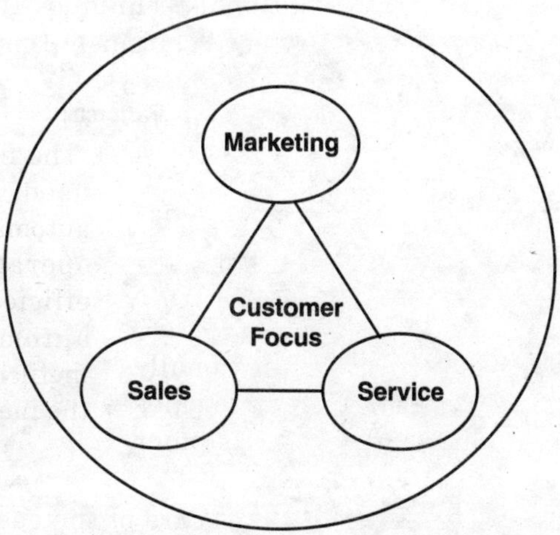

Fig. 13.3. Areas of CRM

13.7. STAGES IN CUSTOMER RELATIONSHIP MANAGEMENT

There are five stages in customer relationship management. These are :

1. **Pre-Relationship Stage :** In the pre-relationship stage, the buyer seeks a new business partner.
2. **Early Stage :** In the early stage, experience is collected between the buyer and seller. But the main problem in this stage is that a great degree of uncertainty and distance is there between the buyer and the seller.

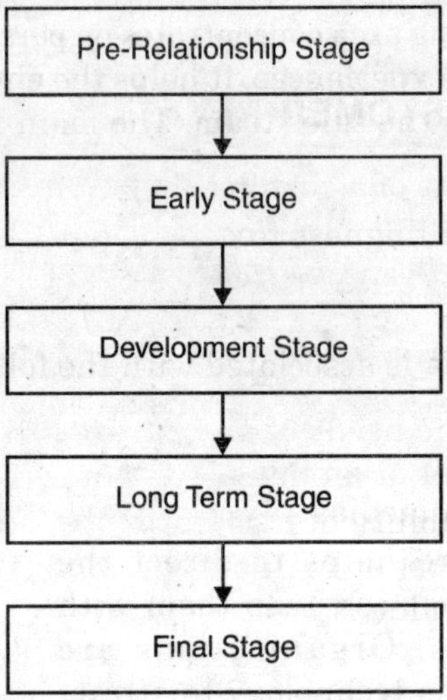

Fig. 13.4. Customer Relationship Management Stages

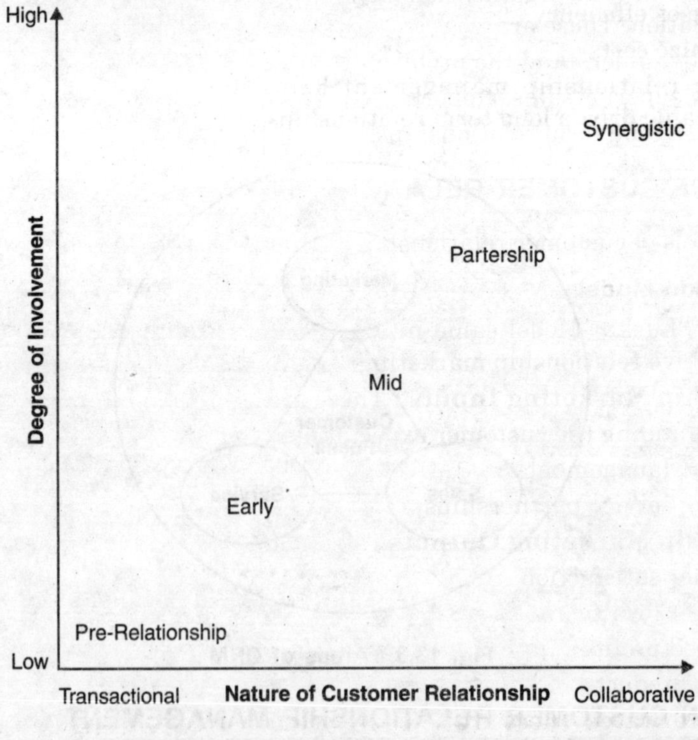

Fig. 13.5. Stages in the development of key account relationship

3. **Development Stage :** In the development stage, the distance is reduced by the social exchange between the buyer and seller. A high degree of communication takes place between the two.
4. **Long Term Stage :** The long term stage is characterised by the company's mutual importance to each another.
5. **Final Stage :** In the final stage, the interaction between the companies becomes grounded.

13.8. DETERMINANTS OF CUSTOMER RELATIONSHIP MANAGEMENT

There are mainly two determinants of a customer relationship management. These are :

1. **Trust :** The trust is the willingness to depend on the :
   - Ability
   - Integrity
   - Motivation

   of one company to serve the needs of another company. This is done in two ways i.e. either in an explicit manner or an implicit manner.
2. **Value :** The value is an ability of a selling organization to satisfy the needs of the customer at either a low cost or high benefits than offered by the competitors of the selling organization. The value is measured in three terms :
   - Temporal
   - Functional and
   - Psychological.

   The salesperson of an organization must do the following operations for the welfare of the organization. These are :
   - Carefully understand the problems and needs of the customer.
   - Provide superior sales support.
   - Establish long term relationships.

13.9. MODELS OF CUSTOMER RELATIONSHIP MANAGEMENT

There are two models of customer relationship management. These are :

1. Evans and Luskin Model

The Evans and Luskin Model came into existence in 1994. The main existence of this model was for effective relationship marketing. The Evans and Luskin Model include :

(*a*) Relationship Marketing Inputs : These are :
- Understanding the customer expectations
- Quality Management
- Building service partnerships

(*b*) Relationship Marketing Outputs : These are :
- Customer satisfaction
- Customer loyalty
- Increased profit
- Quality products

(*c*) Assessment State : These are :
- Customer feedback

• Integration

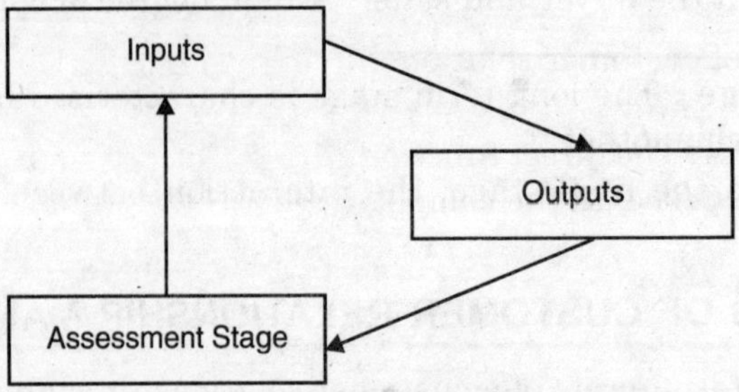

Fig. 13.6. Evans and Luskin Model

2. Brock and Barcklay Model

The Brock and Barcklay model came into existence in 1999. The main existence of this model was for selling partner relationship effectiveness.

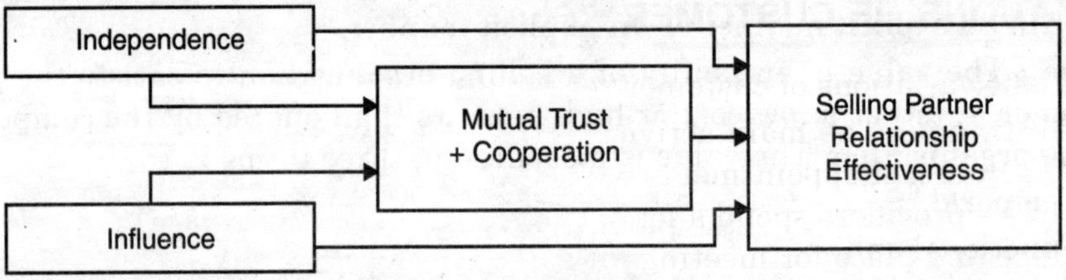

Fig. 13.7. Brock and Barclay Model

13.9.1. Relationship Life Cycle Model

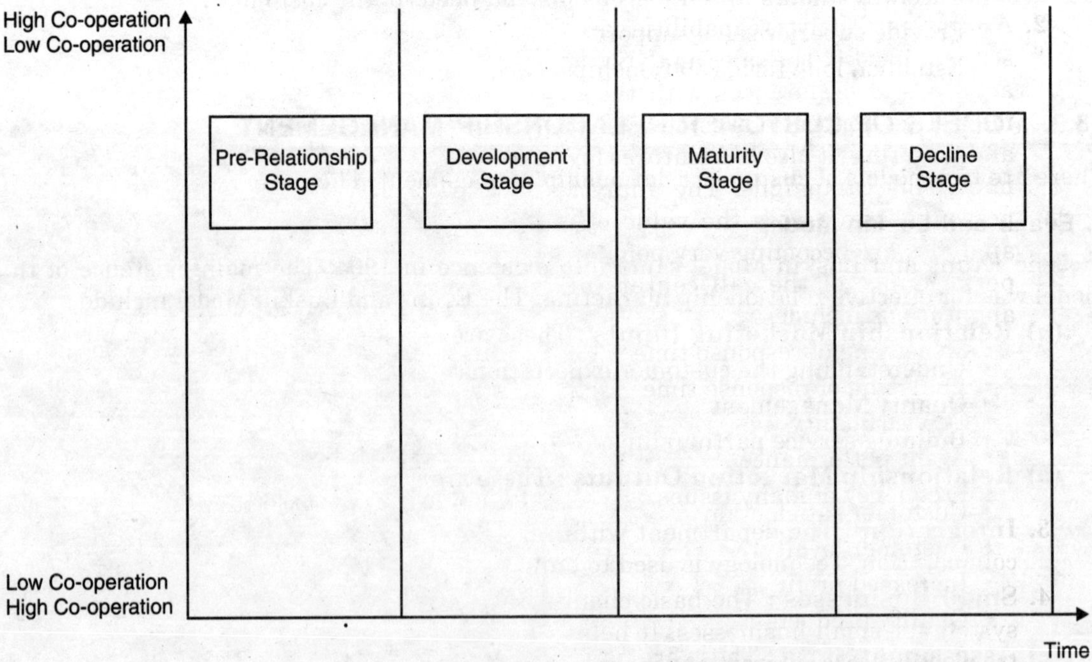

Fig. 13.8. Relationship Life Cycle Model

13.10. FUNCTIONS OF CUSTOMER RELATIONSHIP MANAGEMENT

There are two functions of customer relationship management. These are direct and indirect functions.

1. **Direct Functions :** The direct functions are the basic requirements of a company that are necessary to survive in the competitive market. These are :
   - Profit
   - Volume
   - Safeguard
2. **Indirect Functions :** The indirect functions are the actions that are necessary to convince the customer to participate in the various marketing activities. These are :
   - Innovation
   - Market
   - Access

13.11. VARIATIONS OF CUSTOMER RELATIONSHIP MANAGEMENT

There are various variations of customer relationship management. These are :

1. **Appointment :** The main activity of the customer oriented business is the creation and scheduling of appointments with the customers. The sales, customer support and the service providers spend a lot of time to get in touch with the customers and agree on a time and place for meeting for sales conversation or deliver a customer service. The appointment customer relationship management is a new customer relationship management in which an automated system is used to offer suitable appointment times to the customer through email or web site. This system is integrated with the sales and marketing CRMs to capture and store the interaction with the customers.
2. **Analytics :** Analytic capabilities are linked into applications for sales, marketing and service purpose. The sales analytics let the companies to monitor and understand client actions and preferences with the help of sales forecasting and data quality. The marketing applications come with the predictive analytics to improve segmentation and targeting. It also has features for measuring the effectiveness of online and offline marketing campaigns. The analytics is also used by the marketing and finance personnel's to assess the value of multi-faceted programs as a whole. Nowadays analytics are becoming very popular because they provide greater visibility into the performance of the call centers and other support channels. Support focussed applications includes :
   - Measuring of response time
   - Analysing of response time
   - Service quality
   - Agent performance
   - Frequency of many issues
3. **Integration :** The department within large enterprises tend to function with little collaboration. Technology is used to build the bridge between departments.
4. **Small Businesses :** The basic client service can be provided by a contact manager system for small businesses. It helps the organization and individuals to efficiently track and record interactions including email, documents, jobs, faxes and scheduling. These tools focus more on accounts rather than individual contacts.

5. **Social media :** Social media sites like Twitter, Facebook, LinkedIn and Google plus are increasing the voice of people in the market. They are having a deep effect on the way in which people buy. Customers can research the companies online and discuss about the product online. They can moreover share their opinions and experiences on companies, their products and services. Also the social media is not properly moderated, thus the customers can say their opinions openly, be it be positive or negative. Nowadays many companies are also integrating with the social networking sites.
6. **Non-profit and membership-based :** Systems for non-profit and membership based organizations helps the elements and their involvement in the organizations. The capabilities are :
   - Membership levels
   - Membership directories
   - Volunteering
   - Communication with the individuals.

> **DO YOU KNOW?**
>
> "We've spent the last 30 years focussing on the T in IT, and we'll spend the next 30 years focusing on I."
>
> **–Peter Drucker**

7. **Horizontal vs. Vertical :** The horizontal customer relationship management offer the same non specialized base products across all the industries. They are quite cheaper in the starting and become costly in the future because the company must modify the solution according to a particular industry and business model.
   Vertical customer relationship management offers specialized and specific industry CRM solutions. Vertical solutions are costly in the starting and less costly in the near future because they already include the best practices that are specific to an industry and business model. The vertical CRM vendors focus on a particular industry.
   Most of the vendors offer horizontal CRM solutions.

13.12. STRATEGY OF CUSTOMER RELATIONSHIP MANAGEMENT

A complete and detailed plan is required for large enterprises. It is used to obtain the funds, resources and company wide support that can make the initiative of choosing and implementing the system successfully. There are three major areas where the benefits must be defined. These are :

1. **Processes :** The business systems have many technological components but businesses processes lie at its core. It distributes important information about :
   - Clients
   - Sales
   - Marketing
   - Market trends

   The company must analyse its business workflows and processes before choosing a technology platform. Sometimes, reengineering is required to serve the overall goal of winning and satisfying clients. Planners need to determine the important and relevant client information.
2. **People :** To make an initiative effective, the organization must convince its staff that the new technology will benefit both the employees as well as the clients. It is the responsibility of the senior executives to be strong and visible so that they can clearly support the case for change.
3. **Technology :** The key factors that must be considered in evaluating the technology are :

- Ability to deliver the right data to the right employees
- Ease of adoption
- Ease of use

The platform is selected by a group of executives. These executives understand the software in a good manner. The choosing of application depends upon the size of the company.

13.13. IMPLEMENTATION ISSUES OF CUSTOMER RELATIONSHIP MANAGEMENT

The implementation issues of a customer relationship management are briefly discussed below :

1. **Poor Planning :** The initiative taken by a management can easily fail due to the following reasons :
   - Limitation in choosing a software
   - Limitation in installing the software
   - Limited support from the workforce.
2. **Poor Integration :** For some companies, integration is an important initiative that is taken for :
   - Improving a client facing process
   - Automating the sales
   - Automating the client support channel

   Point solutions offer very little integration to the company's overall strategy. They offer an incomplete client view and unsatisfactory user experiences.

13.14. HELP MENUS

The main issue faced in the implementation and adoption of customer relationship management is the lack of user and technical support in using the system. This is a common problem faced by individuals and the organizations. Thus help menus provide assistance to the users by introducing additional screenshots and other visual and interactive media. There are five major characteristics through which the help menus can be made effective. These are :

1. **Context Specific :** The help menu contains only the information that is relevant to the topic being discussed.
2. **Useful :** The help menu must be comprehensive of all the information that is asked by the user.
3. **Obvious to Invoke :** The user should not find trouble in locating the help menu and to gain access to its contents.
4. **Non-Intrusive :** The help menu should not interfere with the user's primary path of work. It should maintain a distance that allows for its use only when requested.
5. **Easily Available :** The information regarding the help menu must be accessible with only little or no steps.

13.15. ADOPTION RATES OF CUSTOMER RELATIONSHIP MANAGEMENT

There are three recommendations for increasing the adoption rate of customer relationship management. The users should combine the tools in their daily work. These positive recommendations led to an increased usage of the CRM systems and proper implementation. The recommendations are :

1. **Breadcrumb Trail :** The breadcrumb tail provides the user a path that is at the top of a CRM page. It is used to return to the starting point of navigation. It is useful for those users that are not sure on how they got to the current screen in the CRM.
2. **Search Engine Boxes :** The search engines are readily available. With the help of these search engines, the users can immediately get the results. A CRM system that makes use of a search engine provides immediate results within the reach of the user.
3. **Help Option Menu :** The help option menu is an outlet for quick assistance to the users. It makes the customer relationship management software easy to use. The help option menu should be made part of the CRM systems during the development stage.

13.16. BENEFITS OF CUSTOMER RELATIONSHIP MANAGEMENT

The advantages of customer relationship management are :

- Improved quality
- Efficiency is increased
- Costs are reduced
- Customer is provided more attention
- Profits are increased
- Improved planning
- Product development is improved
- Provides decision support
- Sales, service and support staff deals more effectively with the customers because they have a complete view of all customer interactions.
- The management improves since the CRM system makes it quick and easy to produce accurate reports.
- The sales and marketing team starts achieving more business targets because of better tracking and management.
- The data in the CRM systems is quite well organized.
- Due to better forecasting and scheduling, the purchasing improves and cost reduces.

13.17. CHALLENGES FACED BY CUSTOMER RELATIONSHIP MANAGEMENT

The challenges faced by a customer relationship management are listed below :

1. **Complexity :** The tools and workflows are often complex for large business organizations. Initially the tools were limited to simple CRM solutions which focussed on monitoring, recording interactions and communications. But then the tools expanded to deal tracking, territories, opportunities and sales process.
2. **Poor usability :** The major challenge faced by the customer relationship management is poor usability. The importance of usability in a system is developed over time. Capability and complexity of the system have resulted in poor usability of the system. A complex computer system can result in an equally complex and non-friendly user interface, thus the system does not work as intended.
3. **Fragmentation :** Poor usability leads to implementations that are fragmented i.e. those initiatives that are isolated by individual departments to address and fulfil their own needs. It leads to loss in financial benefit that is associated with a customer relationship management.
4. **Business Reputation :** Nowadays building and maintaining a strong business reputation has become quite challenging. Previously, the observations and comments

by the customers on the internal fragmentation were only visible to the employees but now they are visible to the entire world. The human response at all the levels of the organization can affect the customer experience in both good and bad ways.

5. **Security Concerns :** The customer's need for privacy and data security should be considered. A close attention should be paid to the laws and regulations. Vendors must assure their clients that their data is not shared with third parties without their permission so that an illegal access can be prevented.

The major problem faced by the developers and users is to find a balance between the ease of use of the CRM interface and security measures. Researches show the importance of proper designs and technology that balance a positive user interface with appropriate security features. A secure CRM system is more usable. A more usable system creates less confusion thus the amount of errors decreases in turn creating a more secure and stable system. Technical writers play an important role in developing management systems that are both secure and easy to use.

13.18. IMPLEMENTATION OF CUSTOMER RELATIONSHIP MANAGEMENT

In the implementation of a customer relationship management, the following factors must be taken into consideration :

- Easy interaction between customers and the company. It helps in providing quick response to the customer's requests and suggestions.
- Easy access to information about the company. This includes content of customization, advantages of the company, benefits to the customers etc. It helps in maintaining mutual trust and respect.
- Large amount of customer information which have been gathered from different channels.
- Customer's information must be updated with time.
- Improvement in customer relationships increases customer loyalty and increases sales.

13.19. EXAMPLES OF CUSTOMER RELATIONSHIP MANAGEMENT

The companies that must adopt customer relationship management are :

- Banking
- Finance
- Insurance
- Airlines
- Hotels
- Telecommunications
- Health care benefit

13.20. TECHNICAL FUNCTIONALITIES OF CUSTOMER RELATIONSHIP MANAGEMENT

Multi communication channel is a major factor of a customer relationship management. It helps the business to be responsive to the needs and desires of the customers. The various communication channels are :

- **Web channel :** The web channel supports :
  - Customer service activities
  - Customer account management activities.

- Customers can verify and browse product descriptions
- Customers can report problems through this channel.

- **Fax channel :** The fax channel is used by the company to generate fax that is sent directly to customers. The customers use the fax channel to submit requests for information and contact.
- **Interactive TV channel :** Cable TV companies interact with the customers through interactive TV channel.
- **IVR (Integrated Voice Response) :** This channel answers all incoming calls from the toll free network and also provides guidance through the use of prompts.
- **E-mail :** Customers can e-mail their service requests and check its status.
- **Direct field sales force channel :** The direct field sales force channel performs activities like :
  - Browsing of product offerings
  - Capturing leads
  - Completing orders
  - Performing high level customer care

  All the activities are done remotely by the agents with the help of this channel.

Summary

- Customer relationship management is the development and maintenance of mutually beneficial long term relationships with strategically significant customers.
- Customer Relationship Management is also known as relationship marketing or customer management.
- Operational CRM, analytical CRM and collaborative CRM are the three types of CRM.
- The operational CRM helps in easy and effective communication with the customers.
- The aim of the channel management tool is to deliver the products and services across multiple channels in an effective, efficient and consistent manner.
- The collaborative CRM allows the company to synchronize and manage efficient and productive interaction across all the communication channels with customers, prospects, partners and internal associates.
- In collaborative CRM, the viewpoint of the customer is taken into account at every transaction level.
- The example of collaborative CRM is the various departments of the company like sales, technical support, marketing etc.
- Sugar CRM, VTiger and daffodil CRM are three types of open source CRMs.
- Sugar CRM is a commercial open source CRM that is suitable for companies of all sizes.
- Vtiger CRM is the most suitable CRM for small and medium business organizations.
- VTiger was originally built from the source code of Sugar CRM. This CRM is based on open source technologies like Linux, Apache, MySQL and PHP.
- The lead management creates leads and then converts them to suitable opportunities. It imports important leads from web, seminars and mails.
- Daffodil CRM is a commercial CRM. It helps to integrate all the aspects of customer life cycle. It helps in automating everyday tasks.

- Traditional marketing techniques are functionally based. They neglect the customer after the sales process is over and fails to meet the customer needs.
- Customer Relationship Management takes care of the customer needs and is one of the most useful tools in the company's success.
- The three areas of the customer relationship management are sales, marketing and service.
- The sales force predicts and proposes real time analysis of the information that is available and then distributes it to the company.
- The marketing area concentrates on personalizing the customer preferences and offer them satisfying experiences.
- The service area is associated with call centers, emails and other communication medias.
- The call center software helps to direct the customer to the agent that can help them with their current problems.
- Pre-relationship stage, early stage, development, long term and final are the five stages of a customer relationship management.
- In the pre-relationship stage, the buyer seeks a new business partner.
- In the early stage, experience is collected between the buyer and seller. But the main problem in this stage is that a great degree of uncertainty and distance is there between the buyer and the seller.
- In the development stage, the distance is reduced by the social exchange between the buyer and seller. A high degree of communication takes place between the two.
- The long term stage is characterised by the company's mutual importance to each another.
- In the final stage, an interaction between the companies becomes grounded.
- Trust and value are the two determinants of a customer relationship management.
- The value is an ability of a selling organization to satisfy the needs of the customer at either a low cost or high benefits than offered by the competitors of the selling organization.
- The Evans and Luskin Model came into existence in 1994. The main existence of this model was for effective relationship marketing.
- The Brock and Barcklay model came into existence in 1999. The main existence of this model was for selling partner relationship effectiveness.
- Pre-relationship, development, maturity and decline stage are the four stages of the relationship life cycle of a customer relationship management.
- Direct and indirect are the two functions of a customer relationship management.
- The direct functions are the basic requirements of a company that are necessary to survive in the competitive market.
- The indirect functions are the actions that are necessary to convince the customer to participate in the various marketing activities.
- The appointment customer relationship management is a new customer relationship management in which an automated system is used to offer suitable appointment times to the customer through email or web site.
- Analytic capabilities are linked into applications for sales, marketing and service purpose. The sales analytics let the companies to monitor and understand client actions and preferences with the help of sales forecasting and data quality.
- Social media sites like Twitter, Facebook, LinkedIn and Google plus are increasing the voice of people in the market. They are having a deep effect on the way in which people buy.
- Systems for non-profit and membership based organizations helps the elements and their involvement in the organizations.

- The horizontal customer relationship management offer the same non specialized base products across all the industries.
- Vertical customer relationship management offers specialized and specific industry CRM solutions.
- Poor planning and integration are the two implementation issues faced by a customer relationship management.
- Help menus provide assistance to the users by introducing additional screenshots and other visual and interactive media.
- The breadcrumb tail provides the user a path that is at the top of a CRM page.
- The search engines are readily available. With the help of these search engines, the users can immediately get the results.
- Complexity, poor usability, fragmentation, business reputation, security concerns are the major problems faced by a customer relationship management.
- Banking, finance, insurance, airlines, hotels, telecommunications and health care benefits are the examples of customer relationship management.

Short Question Answers

1. Define customer relationship management.

Ans. Customer relationship management is the development and maintenance of mutually beneficial long term relationships with strategically significant customers. Customer Relationship Management is also known as relationship marketing or customer management.

2. What are the three types of customer relationship management?

Ans. Operational, analytical and collaborative are the three types of customer relationship management.

3. What is operational CRM?

Ans. The operational CRM helps in easy and effective communication with the customers. In this type of CRM, various tools are used which are known as contact management tools.

4. What is the benefit of contact management tools?

Ans. The major benefits of contact management tools are :

- Reduces the cost
- Increases efficiency
- Makes use of media based communication channels
- Provides consistent interface across all the communication channels.

5. What are channel management tools?

Ans. Channel management tools aims to understand how the customer interacts with the company. The aim of the channel management tool is to deliver the products and services across multiple channels in an effective, efficient and consistent manner.

6. What are the various processes managed by the channel management tools?

Ans. Campaign management, sales management, service management and complaint management are the various processes that are managed by channel management tools.

7. What are the components of analytical CRM?

Ans. Pattern discovery component, product analysis component, customer analysis component, multitude component, sorting and customer fractionation component and customer value evaluation component are the components of analytical CRM.

8. What are the examples of analytical CRMs?

Ans. Data warehousing, online analytical processing (OLAP) and data mining systems are the examples of analytical CRM.

9. What is collaborative CRM?

Ans. The collaborative CRM allows the company to synchronize and manage efficient and productive interaction across all the communication channels with customers, prospects, partners and internal associates.

10. What are the examples of collaborative CRM?

Ans. The examples of collaborative CRM are the various departments of the company like sales, technical support, marketing etc.

11. What are the three open source customer relationship management?

Ans. Sugar, Vtiger and daffodil are the three open source customer relationship management.

12. What is sugar CRM?

Ans. Sugar CRM is a commercial open source CRM that is suitable for companies of all sizes. The version 3.0 is known as sugar suite.

13. What are the various functions of sugar CRM?

Ans. Sales forecasting, campaign and project management, email marketing, real time tracking, document management and employee directory are the various functions of sugar CRM.

14. What is sales forecasting?

Ans. Sales forecasting helps the sales manager to calculate potential sales and the associated revenues. It provides a better control over performance, commitment, achieving goals and planning resources.

15. What is campaign management?

Ans. The campaign management tool allows users to perform the following functions :

- Organize the marketing activities
- Manage and track marketing efforts on leads, contacts and opportunities.

16. What is project management?

Ans. The project management tool enables the users to :

- Define steps
- Establish timelines
- Assign work to members on any project that is useful to the end users.

17. What is email marketing?

Ans. The marketing team performs the following operations :

- Mail templates
- Listing of receivers
- Sending personalized email templates as per the schedule.

18. What is real time tracking?

Ans. The activities for real time tracking includes calendaring call scheduling, planning meetings and note tracking.

19. What is document management?

Ans. The document management includes capabilities like managing sales documentation and customer reports.

20. What are the components of employee directory?

Ans. First name, last name, role, department and contact information are the components of employee directory.

21. What is Vtiger CRM?

Ans. Vtiger CRM is the most suitable CRM for small and medium business organizations. It was originally built from the source code of Sugar CRM. This CRM is based on open source technologies like Linux, Apache, MySQL and PHP.

22. What are the features of Vtiger CRM?

Ans. Lead management, opportunity and contact management, account management, inventory and activity management and security management are the features of Vtiger CRM.

23. What is lead management?

Ans. The lead management creates leads and then converts them to suitable opportunities. It imports important leads from Web, seminars and mails.

24. What is opportunity and contact management?

Ans. The user can track all the sales opportunities. These opportunities are then associated with accounts, activities, contacts and other modules. The hierarchy of contacts is maintained so that the customers can be dealt in a proper way.

25. What are the operations of account management?

Ans. The user can perform the following operations :

- Import accounts
- Export accounts
- Specify parent-child relationship between accounts
- Track all the accounts
- Attach customer specific documents for quick reference

26. What is activity and inventory management?

Ans. The activity and inventory management helps the user in performing the following functions :

- Add important customer related emails
- Store all the details of customer meetings and calls

27. What is security management?

Ans. The security management feature helps the organization in defining the role of team members and giving them access as per their role.

28. What is daffodil CRM?

Ans. Daffodil CRM is a commercial CRM. It helps to integrate all the aspects of customer life cycle. It helps in automating everyday tasks.

29. What are the features of daffodil CRM?

Ans. Printable reports and export facility, automatic mail attachment, lead management and campaign facility are the features of daffodil CRM.

30. What is printable reports and export facility?

Ans. The daffodil CRM has facilities of :

- Export to excel
- Printable report facilities

The view facility helps to generate different views of data. The users can evaluate the performance of their overall business, sales team and products over a period of time.

31. What is automatic mail attachment?

Ans. Mails can be attached automatically to the daffodil CRM. Manual selection and sending is not required.

32. What is campaign facility?

Ans. Campaign facility enables mass mailings to all or selected list of customers. Email templates may or may not be used for this purpose.

33. What is the difference between customer relationship management and traditional techniques?

Ans. Traditional marketing techniques are functionally based. They neglect the customer after the sales process is over and fails to meet the customer needs. Customer Relationship Management takes care of the customer needs and is one of the most useful tools in the company's success.

34. What are the customer relationship management areas?

Ans. Sales, marketing and value are the three areas of customer relationship management.

35. Explain sales area?

Ans. The sales force predicts and proposes real time analysis of the information that is available and then distributes it to the company. It makes use of the software to reorganize all the phases of the sales process.

36. Explain marketing area?

Ans. The marketing area concentrates on personalizing the customer preferences and offer them satisfying experiences. It helps the enterprise to identify and target the possible clients and helps the sales team.

37. What are the five stages of a customer relationship management?

Ans. Pre-relationship stage, early stage, development, long term and final are the five stages of a customer relationship management.

38. Define pre-relationship stage.

Ans. In the pre-relationship stage, the buyer seeks a new business partner.

39. Define early stage.

Ans. In the early stage, experience is collected between the buyer and seller. But the main problem in this stage is that a great degree of uncertainty and distance is there between the buyer and the seller.

40. Define development stage.

Ans. In the development stage, the distance is reduced by the social exchange between the buyer and seller. A high degree of communication takes place between the two.

41. Define long term stage.

Ans. The long term stage is characterised by the company's mutual importance to each another.

42. Define final stage.

Ans. In the final stage, the interaction between the companies becomes grounded.

43. What are the two determinants of a customer relationship management?

Ans. Trust and value are the two determinants of a customer relationship management.

44. What is trust?

Ans. The trust is the willingness to depend on the ability, integrity and motivation of one company to serve the needs of another company. This is done in two ways i.e. either in an explicit manner or an implicit manner.

45. What is value?

Ans. The value is an ability of a selling organization to satisfy the needs of the customer at either a low cost or high benefits than offered by the competitors of the selling organization. The value is measured in three terms i.e. temporal, functional and psychological.

46. What is the responsibility of the salesperson of an organization?

Ans. The salesperson of an organization must do the following operations for the welfare of the organization. These are :

- Carefully understand the problems and needs of the customer.
- Provide superior sales support.
- Establish long term relationships.

47. What are the twc models of customer relationship management?

Ans. Evans and Luskin Model and Brock and Barcklay Model are the two models of customer relationship management.

48. What are the components of Evans and Luskin model?

Ans. Relationship marketing inputs, relationship marketing outputs and assessment state are the three components of Evans and Luskin model.

49. What are the relationship marketing inputs?

Ans. Understanding the customer expectations, quality management and building service partnerships are the relationship marketing inputs.

50. What are the relationship marketing outputs?

Ans. Customer satisfaction, customer loyalty, increased profit and quality products are the relationship marketing outputs.

51. What is assessment state?

Ans. Customer feedback and integration are the components of assessment state.

52. What are the four stages of the relationship life cycle of a customer relationship management?

Ans. Pre-relationship, development, maturity and decline stage are the four stages of the relationship life cycle of a customer relationship management.

53. What are the two functions of a customer relationship management?

Ans. Direct and indirect are the two functions of a customer relationship management.

54. What is direct function?

Ans. The direct functions are the basic requirements of a company that are necessary to survive in the competitive market. These are profit, volume and safeguard.

55. What is indirect function?

Ans. The indirect functions are the actions that are necessary to convince the customer to participate in the various marketing activities. Innovation, market and access are the three indirect functions.

56. What is appointment customer relationship management?

Ans. The appointment customer relationship management is a new customer relationship management in which an automated system is used to offer suitable appointment times to the customer through email or web site.

57. What are the variations of customer relationship management?

Ans. Appointment, analytics, integration, small business, social media, non-profit and membership based and horizontal vs. vertical are the variations of customer relationship management.

58. What is analytics?

Ans. Analytic capabilities are linked into applications for sales, marketing and service purpose. The sales analytics let the companies to monitor and understand client actions and preferences with the help of sales forecasting and data quality.

59. What is social media?

Ans. Social media sites like Twitter, Facebook, LinkedIn and Google plus are increasing the voice of people in the market. They are having a deep effects on the way in which people buy.

60. What is horizontal vs. vertical?

Ans. The horizontal customer relationship management offer the same non specialized base products across all the industries. Vertical customer relationship management offers specialized and specific industry CRM solutions.

61. What are the implementation issues faced by a customer relationship management?

Ans. Poor planning and integration are the two implementation issues faced by a customer relationship management.

62. What is help menu?

Ans. Help menus provide assistance to the users by introducing additional screenshots and other visual and interactive media.

63. What are the strategies of customer relationship management?

Ans. Process, people and technology are the strategies of customer relationship management.

64. What are the five major characteristics through which the help menus can be made effective?

Ans. Context specific, useful, obvious to invoke, easily available and non-intrusive are the five major characteristics through which the help menus can be made effective.

65. What are the major problems faced by a customer relationship management?

Ans. Complexity, poor usability, fragmentation, business reputation, security concerns are the major problems faced by a customer relationship management.

66. What are the examples of customer relationship management?

Ans. Banking, finance, insurance, airlines, hotels, telecommunications and health care benefits are the examples of customer relationship management.

67. What are the recommendations for customer relationship management?

Ans. Breadcrumb trail, search engine and help menus are the recommendations for customer relationship management.

Review Questions

1. Explain customer relationship management in detail.
2. Explain the types of CRM in detail.
3. Explain the open source CRM in detail.
4. What are the goals of CRM?
5. Explain the areas of CRM in detail.
6. Explain the stages of CRM in detail.
7. Explain the determinants of CRM in detail.
8. Explain the models of CRM in detail.
9. Explain the functions of CRM in detail.
10. Explain the variations of CRM in detail.
11. Explain the strategy of CRM in detail.
12. Explain help menu in detail.
13. What are the advantages and disadvantages of CRM?
14. What are the technical functionalities of CRM?
15. Compare traditional marketing techniques and CRM.

Chapter 14

Procurement Management System

"How you gather, manage and use information will determine whether you win or lose."

Inside this Chapter

- Procurement Management Overview
- History of Procurement Management
- Procurement Management Functions
- Types of Procurement Management
- Procurement Management Life Cycle
- Procurement Management Example

14.1. PROCUREMENT MANAGEMENT OVERVIEW

A procurement management process or a procurement process is the method of purchasing goods and services from external suppliers. Obtaining goods and services from external suppliers can be a critical path for many projects. The performance of the supplier reflects the performance of the overall project team. The supplier's performance should be monitored carefully and regularly so that the delivery of the product meets the expectations of the team.

The goods and services should be obtained at the best possible cost to meet the needs of the purchaser in terms of:

- Quality
- Quantity
- Time
- Location

The management process provides procurement procedures. This procedure consists of step-by-step explanation of how to purchase goods from suppliers.

Thus the procurement management process helps you: "**Get What You Have Paid For**".

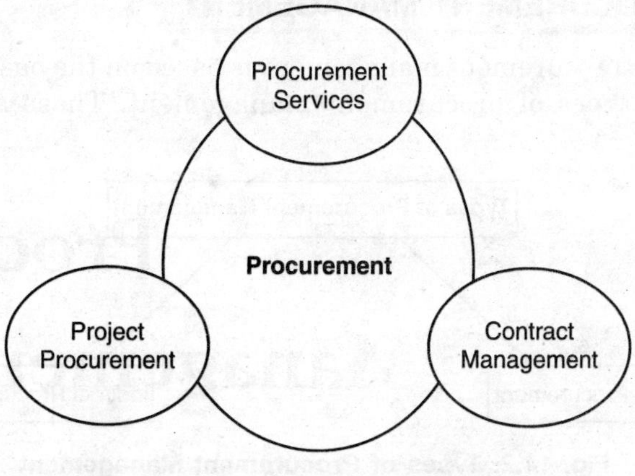

Fig. 14.1. Procurement Management

14.2. HISTORY OF PROCUREMENT MANAGEMENT

Before the World War-1, purchasing of items was considered as clerical. During the era of World War 1 and 2, the function of purchasing increased because great importance was given to raw materials, supplies and services. These operations were required to keep the mines and other factories in operation.

During **1950-1960s**, techniques for performing purchasing functions became more refined. The number of trained professionals also increased.

During **1970-1980s**, more importance was given to the purchasing strategy. More focus was given on the techniques to obtain the items from the supplier at genuine prices.

In **1990s**, the procurement process became more integrated. Supply chain management software's were developed.

In **2000s**, the leader of the procurement function within many enterprises was established through a Chief Procurement Officer. The Chief Procurement Officer is also known as the Head of Procurement. The officers were considered as important business leaders and have more operational responsibility.

14.3. PROCUREMENT MANAGEMENT FUNCTIONS

The procurement management process consists of the following functions :

- Identify the goods and services that are to be procured
- Managing the orders
- Managing the receipts
- Review of items from the suppliers
- Approval of items from suppliers
- Managing the relationship with the supplier
- Ensure that a high level of service is received
- Review the supplier performance against the contract
- Identify and resolve supplier performance issues
- Approve the supplier payments
- Communicate the status to the management

DO YOU KNOW?

I've always wanted to own and control the primary technology in everything we do.

— Steve Jobs

14.4. TYPES OF PROCUREMENT MANAGEMENT

The classification of the procurement management is based on the basis of the consumption of goods. There are two types of procurement management. These are direct and indirect procurement.

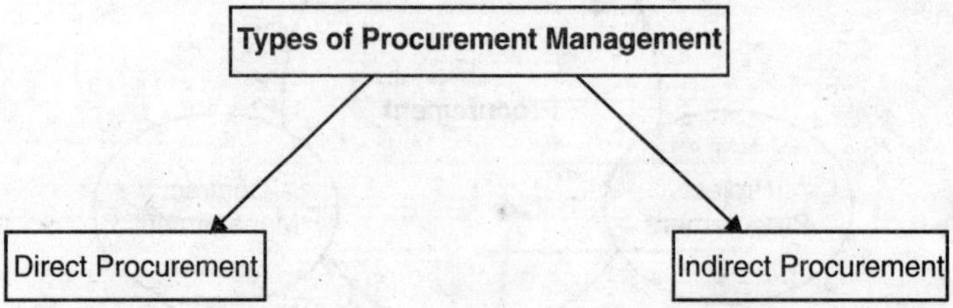

Fig. 14.2. Types of Procurement Management

Direct Procurement : The direct procurement is production related. It takes place during the manufacturing process only. It consists of the items that are part of the finished product. For example : raw material, components and parts. It affects the production process of the manufacturing firms.

Features of Direct Procurement

The features of direct procurement are :

- Large quantity
- High frequency
- Industry specific value
- Operational in nature

For Example: Crude oil in petroleum industry.

Indirect Procurement : The indirect procurement is non-production related. It deals with the operating resources. Operating resources are the resources that are purchased by a company to enable its operations.

Features of Indirect Procurement

The features of indirect procurement are :

- Low quantity
- High frequency
- Low value
- Tactical in nature

For Example : Lubricants and spare parts

14.5. PROCUREMENT MANAGEMENT LIFE CYCLE

The procurement management life cycle consists of eight steps. These are :

1. **Information Gathering :** It is the responsibility of the sales and marketing team to gather the appropriate suppliers who can fulfil specific requirements.
2. **Supplier Contract :** When the supplier has been identified, following operations must be performed :
   - Request for Quotation
   - Request for Proposal
   - Request for Information

- Request for Tender

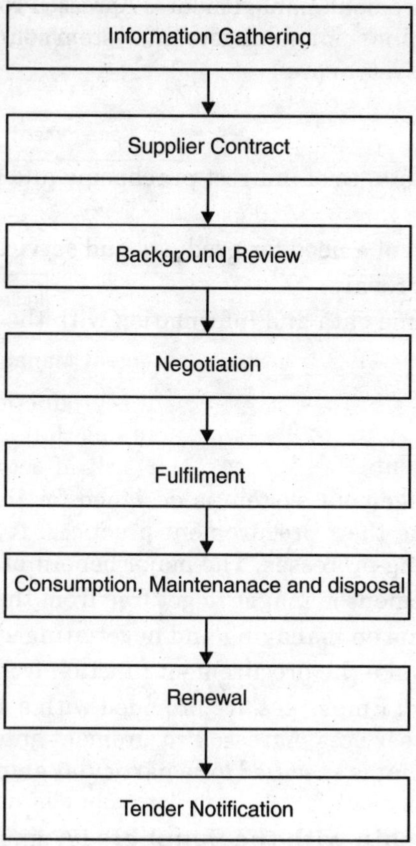

Fig. 14.3. Procurement Management Life-Cycle

3. **Background Review :** In the background review, reference for product and service quality is consulted. Requirements for services like installation, maintenance and warranty is investigated.
4. **Negotiation :** Negotiation for price and delivery schedule is done.
5. **Fulfilment :** The following operations are done :
   - Shipment delivery
   - Payment
   - Installation
   - Training
6. **Consumption, maintenance and disposal :** In this phase, the company evaluates the performance of the services.
7. **Renewal :** When the services are consumed, the contract expires. If it has to be reordered, the company must determine whether to continue with the sample supplier or some another supplier.
8. **Tender Notification :** Many institutions make use of the notification service in order to raise the competition for the chosen opportunity. These systems can either be direct with the help of their re-tendering software or as re-packaged notification from an external notification company.

14.6. PROCUREMENT MANAGEMENT EXAMPLE

One of the examples of procurement management is Agresso Procurement. Agresso procurement is a part of fully integrated financial management system solution. The main functions of this procurement management system are :

- Support
- Automate and
- Standardize all the stages of indirect purchasing and procurement cycle.

It includes the following :

- Initial identification of a need for products and services
- Expenditure for approval
- Exchange of electronic data and information with the suppliers
- Delivery of goods
- Inspection of goods
- Invoice matching
- Settlement of payment

The procurement management system is designed for those organizations that want to control costs and streamline their procurement practices. It is suitable for both centralized and decentralized purchasing processes. The major benefit of using this system is that :

- It helps the procurement manager to get free from the day to day tasks.
- It helps them to focus on managing and negotiating stronger supplier relationships.

Their major role is to make the procurement function from a cost to profit centre.

Single Point of Entry : Employees are provided with a single and easy to use screen for the procurement of goods or services. Agresso Procurement automatically identifies and applies the correct buying process that is specified for a particular employee or role. It helps in making the purchasing cycle visible.

Strengthens relationship with the Suppliers : A single and centralized procurement management system provides :

- Clear
- Complete and
- Accurate analysis of the suppliers of an organization.

It helps in negotiating better terms and prices for the goods.

Purchasing Information : There is a purchase entry screen that is available so that the users can choose whatever they want to purchase. This screen provides :

- Accurate product
- Information related to price

From the portals and websites of the suppliers. This helps in :

- Reducing the overhead
- Better accuracy
- Better control

Authorization and Budget Control : The tools of Agresso provides authorization and approval to match the requirements of each organization or department. It helps in making the entire process quite simple. For example : It automates the transfer process, thus eliminating the manual work. It helps in reducing time and cost.

Summary

- A procurement management process or a procurement process is the method of purchasing goods and services from external suppliers.
- The management process provides procurement procedures. This procedure consists of step-by-step explanation of how to purchase goods from suppliers
- There are two types of procurement management. These are direct and indirect procurement.
- The direct procurement is production related. It takes place during the manufacturing process only.
- The indirect procurement is non-production related. It deals with the operating resources.
- The procurement management system is designed for those organizations that want to control costs and streamline their procurement practices.
- Procurement practice is suitable for both centralized and decentralized purchasing processes.

Short Question Answers

1. Define procurement management.

Ans. A procurement management process or a procurement process is the method of purchasing goods and services from external suppliers.

2. State the procurement management functions.

Ans. The procurement management process consists of the following functions :

- Identify the gocds and services that are to be procured
- Managing the orders
- Managing the receipts
- Review of items from the suppliers
- Approval of items from suppliers
- Managing of the relationship with the supplier
- Ensure that a high level of service is received
- Review the supplier performance against the contract
- Identify and resolve supplier performance issues
- Approve the supplier payments
- Communicate the status to the management

3. What are the types of procurement management?

Ans. Direct and indirect are the two types of procurement management.

4. What is direct procurement?

Ans. The direct procurement is production related. It takes place during the manufacturing process only. It consists of the items that are part of the finished product. For example: raw material, components and parts. It affects the production process of the manufacturing firms.

5. What is indirect procurement?

Ans. The indirect procurement is non-production related. It deals with the operating resources. Operating resources are the resources that are purchased by a company to enable its operations.

6. What are the features of direct procurement?

Ans. The features of direct procurement are :

- Large quantity
- High frequency
- Industry specific value
- Operational in nature

7. What are the features of indirect procurement?

Ans. The features of indirect procurement are :

- Low quantity
- High frequency
- Low value
- Tactical in nature

8. What are the phases of procurement management lifecycle?

Ans. Information gathering, supplier contract, background review, negotiation, fulfilment, consumption, maintenance and disposal, renewal and tender notification are the phases of procurement management lifecycle.

9. What are the main functions of the Agresso procurement management?

Ans. The main functions of this procurement management system are :

- Support
- Automate and
- Standardize all the stages of indirect purchasing and procurement cycle.

10. What is the benefit of the Agresso procurement management?

Ans. The benefit of using this system is that :

- It helps the procurement manager to get free from the day to day tasks.
- It helps them to focus on managing and negotiating stronger supplier relationships

Review Questions

1. Explain procurement management in detail.
2. Explain the history of procurement management.
3. What are the functions of procurement management?
4. Explain the types of procurement management in detail.
5. Explain the lifecycle of procurement management in detail.
6. Explain the example of procurement management in detail.

Previous Year Question Papers (Unsolved)

Note : Two questions will be set from each of the four sections. The students have to attempt first common question, which is compulsory, and one question from each of the four sections. Thus students will have to attempt 5 questions out of 9 questions.

Previous Year Question Papers (Unsolved)

PREVIOUS YEAR QUESTION PAPER-1

1. **Attempt all the questions in brief :**
   (*a*) What is the hardware and software integration of an information system?
   (*b*) What is a system?
   (*c*) What are the elements of a business organization subsystem?
   (*d*) Explain multidimensional structure.
   (*e*) What is the input of the conceptual system design?
   (*f*) What are the components of management control?
   (*g*) What is trade-off criterion?
   (*h*) Explain internal records.
2. Explain information system in detail.
3. Write in detail about the software, hardware and communication technology used in MIS. How the concept of data mining and warehousing is playing a crucial role?
4. Explain procurement management in detail.
5. Explain decision support system.
6. What is problem definition? Why is it required? Determine system objectives and system constraints.
7. What are the steps in planning implementation of MIS?
8. Write short notes on :
   - Detailed system documentation
   - Role of manager in MIS
9. Explain management information and system approach.

❑❑❑

PREVIOUS YEAR QUESTION PAPER-2

1. **Attempt all the questions in brief :**
   (*a*) What is the first view of an information system?
   (*b*) Explain the management concept.
   (*c*) Who is a business information manager?
   (*d*) Explain business process re-engineering.
   (*e*) Explain program logic.
   (*f*) Define constraint.
   (*g*) What is the role of manager in the documentation process?
   (*h*) What is dominant criterion?
2. (*a*) Explain management information system in detail.
   (*b*) Briefly discuss system view of business
3. Explain conceptual design system in detail.
4. What is meant by detailed system design? What are its main aims?
5. How will you evaluate, control and maintain the system?
6. Explain fundamental weakness of MIS in detail.
7. Explain enterprise resource planning.
8. Explain procurement management.
9. (*a*) Briefly discuss project management of MIS detailed design.
   (*b*) Discuss the need of system testing and how is it performed.

❑❑❑

PREVIOUS YEAR QUESTION PAPER-3

1. **Attempt all the questions in brief :**
   (*a*) What are the three types of system flowcharts?
   (*b*) Explain system oriented flowcharts?
   (*c*) What is a database?
   (*d*) Who is a system analyst?
   (*e*) State the ERP vendors.
   (*f*) Explain the evolution of supply chain management.
   (*g*) What are the four supply chain elements?
   (*h*) Explain push strategy.
2. How information needs are determined? How the information is collected?
3. (*a*) What are the activities involved in system implementation? What are the problems encountered?
   (*b*) What do you mean by alternative conceptual design?
4. Write short notes on :
   - Decision support system
   - Soft spots in MIS planning
   - Software and hardware determination for system
5. Explain manager's overview.
6. Explain customer relationship management.
7. What are the steps in planning implementation of MIS?
8. What are the objectives of conceptual system design? How is the conceptual design report prepared?
9. Explain the following:
   - Maintenance of MIS
   - Establish system objectives
   - Propose an organization to operate the system

❑❑❑

Previous Year Question Papers (Solved)

Note : Two questions will be set from each of the four sections. The students have to attempt first common question, which is compulsory, and one question from each of the four sections. Thus students will have to attempt 5 questions out of 9 questions.

Previous Year Question Papers (Solved)

PREVIOUS YEAR QUESTION PAPER-1

Q.1. Attempt all the questions in brief :

(*a*) What are the operations performed by a manager?

Ans. Planning, organizing and controlling are the three management activities performed by the manager.

(*b*) What is tax efficient supply chain management?

Ans. Tax Efficient Supply Chain Management is basically a business model that considers the effect of tax in the design and implementation of the supply chain management. The business organizations that are in different countries should pay different tax rates in different countries. It is thus used as a method of gaining more profit for company which owns global supply chain.

(*c*) What is operational CRM?

Ans. The operational CRM helps in easy and effective communication with the customers. In this type of CRM, various tools are used known as contact management tools.

(*d*) What is indirect procurement?

Ans. The indirect procurement is non-production related. It deals with the operating resources. Operating resources are the resources that are purchased by a company to enable its operations.

(*e*) Explain task force meetings.

Ans. The task force meeting consists of the managers and key specialists. It is an important process.

(*f*) Explain forms oriented charts.

Ans. Form oriented charts are the ones that identify the forms that are used in the communication process. It traces the flow of all the forms through the organization.

(*g*) Explain the process of interview.

Ans. Interview process of managers and personnel's is a method of identifying the possible sources of information. It helps in analyzing the existing information system. The interview process helps in obtaining secure information.

(*h*) Explain the supplier constraint.

Ans. The suppliers must also be taken into account during the designing process of an information system.

Q.2. Explain decision support system.

Ans. A decision support system is basically an interactive computer based information system that supports the decision making activities of an organization. It serves the management level, operations level and planning level.

The decision support system includes knowledge based system. It helps the decision makers to compile useful information from raw data, documents, business models and personal knowledge.

Classification of Decision Support System : The decision support system can be classified into five parts. These are :

(*a*) **Communication-driven decision support system :** This type of DSS supports more than one person that is working on a shared task. For example : Groove, Microsoft's NetMeeting.

(*b*) **Data-driven decision support system :** It is also known as data oriented decision support system. It stresses on the access and manipulation of the time series of internal and external data of the company.

(*c*) **Document-driven decision support system :** This type of DSS manages, retrieves and manipulates unstructured information in different electronic formats.

(*d*) **Knowledge-driven decision support system :** This DSS provides specialized expertise in problem solving. It is stored as facts, rules and procedures.

(*e*) **Model-driven decision support system :** This type of DSS makes use of data and parameters that are provided by the users to help in the decision making process.

Components of Decision Support System : There are three fundamental components of decision support system. These are :

- **Database :** The database is also known as knowledge base.
- **Model :** The model is for either decision context or the user criteria.
- **User interface :** Users are an important part of the DSS architecture.

Decision Support System Framework : The framework of decision support system includes :

- People
- Technology
- Development approach

Q.3. Explain the role of following technologies in the development of MIS.

(*a*) DBMS

Ans. A database is a collection of data records, files and objects. A data base management system is software package that :

- Creates the database
- Maintains the database
- Use the database

It can be accessed by multiple users and multiple application programmers at the same time.

Components of a Database : The components of a database are :

- **DBMS Engine :** It accepts requests from other database systems and converts them into physical parts. It accesses the database and data dictionary.
- **Data definition subsystem :** It helps the user to create the data dictionary and define the structure of the files.

- **Data manipulation subsystem :** As the name suggests, it helps the user to manipulate the system. It helps the user to add, modify and delete the information from the database.
- **Application generation subsystem :** It helps the user to develop applications. The user should perform a detailed series of tasks to process a transaction.
- **Data administration subsystem :** It helps the user to manage the entire database by providing facilities like backup and recovery, security management etc.

Features of Database : The features of database are listed below :

(*a*) Support large amount of data
(*b*) Data Sharing
(*c*) Data Concurrency
(*d*) Data Locking
(*e*) Data Security
(*f*) Data Integrity
(*g*) Fault Tolerance
(*h*) Fault Recovery
(*i*) Language Support

Database Models : There are five database models which are listed in a detailed manner below :

(*a*) Hierarchical Model : The hierarchical database model is a model in which the data is organized in a tree like form. The hierarchical model is a simple model. It represents information using one to many relationships. **For example :** Storing geographical information, mobile etc.

(*b*) Network Structure : The network structure is a database model of representing objects and their relationships. It is a flexible model. The network structure implements many to many relationships.

(*c*) Relational Structure : Relational structure is a database model that makes use of two dimensional rows and columns to store the data. It is a complex method and is not suitable for the end user. The relational structure is used by the mainframe, midrange and microcomputer systems.

(*d*) Multidimensional Structure : The multidimensional structure is very similar to the relational model. The multidimensional structure is quite easy to maintain and understand. The records are stored as fundamental attributes.

(*e*) Object-oriented structure : The object oriented structure is a popular for the multimedia. It has an ability to handle graphics, pictures, voice and text. It is designed to work with the object oriented programming languages like java.

Advantages of Database : The advantages of database are :

- It keeps a check on the data redundancy.
- It ensures data integrity.
- It ensures that the data can be shared across all applications.
- It contains a backup and recovery procedure of data.
- It provides better services to the users.
- Improved flexibility.
- Cost of developing and maintaining the system is low.
- Improved security.

Disadvantages of Database : The disadvantages of using database are :

- Implementation of a database can be expensive.
- Implementation of a database can be time consuming.
- Even with the provision of high security, it is possible for unauthorized users to access the database.

(*b*) Data-ware housing.

Ans. Data warehouse is a repository of information that is organized so that it can be accessed easily for analytical processing activities. Data warehouse is a collection of information that is obtained from either :

- Operational systems
- External data sources

Purpose of Data Warehousing : The purpose of data warehousing is :

- **Realize the value of data :** The realization of data is done by methods of reporting and analysing.
- Better decision making
- Convert data into information
- Create competition

Components of Data Warehousing : There are three components of data warehousing. These are :

(*a*) **Staging Area :** Staging area is a repository where the transaction data can be transferred for and in the data warehouse.

(*b*) **Data Mart :** The data warehouse is the union of a set of data marts.

(*c*) **Operational Data Store :** The operational data store is modelled to support the real time reporting needs.

Characteristics of Data Warehousing : The characteristics of data warehousing are :

(*a*) **Organisation :** The data is organized by subject like customer, vendor etc. It contains information relevant for making decision support only.

(*b*) **Consistency :** Consistency means that data in different database may be encoded differently.

(*c*) **Time variant :** The data is kept for a long time so that it can be used for forecasting purpose.

(*d*) **Non-volatile :** Non-volatile means that the data cannot be updated when it enters the data warehouse.

(*e*) **Relational :** The data warehouse makes use of a relational database.

(*c*) Data mining.

Ans. Data mining is a tool for analysing large amount of data that is present in a data warehouse. It helps in decision making process. It is also known as "Knowledge discovery of database". The goal of the data mining process is to extract knowledge from the existing data and transform into a human understandable format so that it can be used further.

Data Mining Techniques : The most common used techniques in data mining are :

(*a*) **Case-based reasoning :** This uses historical cases to recognize patterns.

(*b*) **Neural computing :** Neural computing is a machine based learning approach. The historical data is examined for pattern recognition. These patterns are used for decision support.

(*c*) **Intelligent agents :** It is an approach for retrieving information from the Internet. A large amount of information is available on the Internet.

(*d*) Other Tools : Many tools like decision trees, genetic algorithms are used in the data mining process.

Applications of data mining : There is a large variety of data mining. These are :

(*a*) Retailing and sales
(*b*) Banking
(*c*) Manufacturing and Production
(*d*) Insurance
(*e*) Computer hardware and software

Capabilities of data mining : There are some advantages of using data mining that are listed below :

(*a*) Automated prediction of trends and behaviours : Data mining automates the process of finding predictive information in large databases. For example : targeted marketing, forecasting bankruptcy.

(*b*) Automated discovery of previous unknown patterns : Data mining identifies previous unknown data in a single step.

(*c*) They analyse massive database in fraction of minutes.

(*d*) Data mining procedure can be easily conducted by non-programmers.

Data mining failure : The reasons for data mining failure are :

- Inappropriate architecture
- Lack of training
- Poor use of technology
- Lack of data quality
- Lack of user input
- Interference of corporate politics

Q.4. Describe the content of conceptual design. Explain how to ensure that :

- **Objectives are clearly stated**
- **System constraints are clearly understood**

Ans. The conceptual system design is an extension of the logical design of the system. The conceptual system design takes into account the various technological factors that are required for achieving the functional requirements of the system. It is also known as feasibility design, gross design or high level design.

The input of the Conceptual System Design is :

- Management information requirements
- Management objectives

The output of the Conceptual System Design is :

- Set of documents that describe the MIS in full detail. It helps the technicians to start their work on the detailed design of the system.

The key tasks that are performed in the Conceptual Designing of the system are :

- Defining the problems in a complete manner.
- Refining the management objectives to set the system objectives.
- Establishing the system constraints.
- Determining the Information needs and requirements.
- Developing the alternatives design and selecting one appropriate one.
- Documenting the conceptual system design phase.

Objectives are clearly stated : The system objectives of an organization must be set. System objectives are expressed in terms of what the manager can do when the information requirements of an organization is fulfilled.

- The setting of system objectives of an organization that cut across all the functional areas is very difficult.
- It is easy to determine the objectives of an operational system that is served by a main system.

The system objectives are viewed by the following people :

(*a*) **Technician :** The technician sees the system objectives of the smaller system as an input to a larger system.

(*b*) **Manager :** It is the responsibility of the manager to define the system objectives in terms of the information needs. They should not define the terms according to the satisfaction of demands.

The system objectives must be stated in terms of :

- Objectives of the department
- Objectives of the group
- Objectives of the function
- Manager to be served

Establish System Constraints : A constraint helps the designer to modify the conditions under which the objectives can be fulfilled. It also helps the designer to consider the limitations that restrict the design. A constraint can have both positive and negative effect.

Classification of Constraints : Constraints can be classified as internal and external constraints.

Internal Constraints : The internal constraints are of five types. These are :

(*a*) **Top Management Support :** If the top management support is not obtained, then the design cannot be implemented. It is the responsibility of the top management to approve and support the design.

This internal constraint influences on the decision of the management on the kind of the system that is supposed to be designed.

(*b*) **Policy Considerations :** Policy considerations should be kept in mind. It helps in setting the limit for objectives. It also helps in modifying an already existing approach to design a system. The flow of information is influenced by the :

- Structure of the organization
- Position of the managers

The organizational policies limit the approach to system design.

(*c*) **Personnel Needs :** The personnel needs are a major limiting factor in the design of Information System and utilization of Information System.

(*d*) **Cost :** Cost is another constraint that is a resource limitation. It should be noted that the cost to achieve the objectives should be compared with the benefits that will be achieved.

(*e*) **Restrictions :** Restriction is the constraint that is made by the managers and designers on the design of an information system. In the designing phase, the manager needs to alter some of the objectives so that the system is fit with other outputs and equipment. The manager will also restrict on the time and effort that is applied to the entire phase.

External Constraints : The external constraints are of three types. These are :

(*a*) **Customer :** Some of the systems like :

- Order entry
- Order billing

Should be designed keeping in mind the needs of the customer. If some outputs from the system are not acceptable, customer constraint may be faced.

(*b*) **Government :** The governmental constraint is of three types :

- Federal
- State
- Local

This type of constraint imposes restrictions on :

- Maintenance of the security of information
- Meeting of certain procedures of record keeping
- Reporting to stockholders

(*c*) **Suppliers :** The suppliers must also be taken into account during the designing process of an information system.

Q.5. Describe various possible pitfalls in MIS design and development?

Ans. Management Information System are risky systems since they aim at changing the organizations. The main problems faced by an organization are :

- New group of stakeholders
- New technology
- Less experienced professionals
- Partly implemented functions
- Over budget
- Delay over implementation

These problems cause losses to an organization and improper use of resources.

Types of Pitfalls : There are four pitfalls in a management information system. These are :

(A) Fundamental Weaknesses : Fundamental weaknesses are faced in an organization only. When a company decides to develop a Management Information System, these fundamental weaknesses are faced.

Review

- **No Management System to build Upon** : The MIS must be built on the top management system. This includes organizational arrangement, structure for proper planning and control, procedure for proper planning and control and clear creation of objectives. The process of management is not performed well due to lack of managerial applications and lack of operational applications.
- **Company Objectives :** The Company without objectives is like a company without a statement of mission and purpose. This lacking of the statement of mission and purpose is a common weakness. This is the weakness caused by the top management of the company due to which the entire employee workforce is left without the benefit of the valuable guidance the objectives provide. The major benefits of setting up of clear objectives are that it helps in decision making and helps in inference purposes.
- **Managerial Participation :** The development of the Management Information System is the responsibility of the management. Managerial participation helps in the

development of a better and effective information system. The management includes top level management and operating line management.

- **Organization of the MIS Functions** : The lack of organization of the MIS functions leads to the system failure.
- **Communication Gap** : In an organization, communication gap is faced at various levels. The computer technician will never speak the language of the management and vice versa. Thus due to the communication gap, design problems arises.
- **People Involved** : There are many key people involved in an organization. The performance of the employees can be classified as good performers, average performers and poor performers.

(B) Soft Spots in Planning : Soft spots in planning are related to the Management Information System that is under planning and development.

Review

- **MIS response to a Business Plan :** The main problems that are faced by the MIS planning are that the personnel's associated with the planning process forgets the original basic purpose of the management. The professionals face difficulties for responding just to the information needs of the main business. One of the difficulties is the temptation to purchase the latest hardware and software techniques. When a change is made to the business plan, the corresponding MIS must be made and changed. Each MIS plan must be a proper response to a business plan.
- **One for one conversion :** The one for one conversion provides an opportunity to upgrade an existing system into a Management Information System. Suppose a technician converts an existing computer system or a manual and converts it without upgrading or changing it.
- **System View** : In a system view, there should be a master plan available with the help of which MIS planning and development can be performed. Without a master plan, a system failure occurs.
- **Setting Project and System Objectives** : System objectives are required for the development of a correct plan. Also, objectives cannot be made until some of the general planning variables are known. Thus both the setting up of planning and objectives goes hand in hand.
- **Facing Constraints** : Constraints are faced on financial definition, system performance, system cost and development schedule.
- **Plan to Sale the MIS** : System designers face the problem of gaining acceptance from the user for whom the system is designed. Thus it should be taken care that the system should be designed in such a way that it is easily accepted by the users. It is the toughest part of the entire designing and implementation phase.
- **Detail Planning** : The MIS developers must plan in detail each phase of the project.

(C) Design Problems : Design problems are the pitfalls faced during the designing phase. These pitfalls are under the control of managers and can be easily solved.

Review

- **Consider Alternatives Designs** : During the early phase of the designing process, many ideas should be considered. It is the responsibility of the designer to provide the various alternative designs with the positive and negative impact of each. As the time passes, the frequency of alternate designs also decreases.

- **Beware the User Interface** : Beware the User Interface is partially a technical problem. The user interface should be according to a user. User should be comfortable with the design or interface of the system. The advantage is that it helps in the conversation among the computer experts and eliminates the conversion between the technician and user.
- **The Real World-The Acid Test** : The business organizations are not research institutes. Thus, the MIS is being implemented to support the firm's main line of business and not to extend the state of art in MIS design.
- **If it Moves-Automate it :** Only some things can be automated and not all.
- **The Computer Obsession :** Computer should not be an obsession in a system. It is a tool and should be used as a tool only.
- **Documentation** : Documentation is a problem that is faced in the planning, designing and implementation phase. The following items must be documented in detail i.e. plans, project and system objective, specification of functions, specification of performance, user interface specifications, user instruction manual, reference manual and maintenance guideline.

(D) Implementation Problems : Various types of implementation problems are faced in the development of a management information system.

Review

- **Hardware Perspective :** In the hardware perspective, the managers show off their computer centers, disk space and the millions of instructions the processor executes per second. Nowadays the hardware items are becoming cheap and are of little concern in the MIS purchase design.
- **Software Development** : The software development pitfall includes two pitfalls. These are hardware related and personnel related. The hardware performs the instructions. Therefore wrong instructions lead to wrong results.
- **Test it and Test It Again** : A major error that is seen is improper planning that is done to perform the testing. Testing must be performed at the functional level, component level and system level. Problems are experienced at each level and they must be fixed.
- **Controlling the MIS Project** : Controlling the MIS Project is the one of the four basic management functions. It also remains one of the major causes of MIS development project failure.

Q.6. Explain customer relationship management.

Ans. Customer relationship management is the development and maintenance of mutually beneficial long term relationships with strategically significant customers. It is an IT improved value process which identifies, develops, integrates and focusses on the various competencies of the firm to the "voice of customer" in order to develop a long term customer value.

Customer Relationship Management is also known as relationship marketing or customer management. Customer relationship management captures, analyses and distributes all relevant data from the customer to everyone in an organization. This distribution of information helps the organization in meeting the customer, product and service needs.

Types of Customer Relationship Management : There are three types of customer relationship management. These are :

1. **Operational CRM :** The operational CRM helps in easy and effective communication with the customers. In this type of CRM, various tools are used known as contact management tools. The major benefits of these tools are :

- Reduces the cost
- Increases efficiency
- Makes use of media based communication channels
- Provides consistent interface across all the communication channels.

Here relevant customer data is collected and displayed at appropriate customer touch points. This is known as **customer master data**. There is another set of data also which consists of information like hobbies, interests, details about children etc. This set of data is known as **soft customer data**. The example of operational CRM is banks.

Channel management tools aims to understand how the customer interacts with the company. The aim of the channel management tool is to deliver the products and services across multiple channels in an effective, efficient and consistent manner. The various processes that are managed by the channel management tools are :

- Campaign management
- Sales management
- Service management
- Complaint management

2. **Analytical CRM :** Analytical CRM is composed of the following :

- Pattern discovery component
- Product analysis component
- Customer analysis component
- Multitude component
- Sorting and customer fractionation component
- Customer value evaluation component

The analytical solutions are an integrated view of :

- Customers across all the channels
- Customers across all the applications
- Campaign performance analysis
- Cross selling
- Customer profit analysis

The analytical CRM solution helps in answering the following questions :

- Who are the best customers of an organization?
- Which customers an organization can loose?
- How to bind the customers with the organization?
- How to attract new customers?
- How to improve the profitability of customers?

The examples of analytical CRMs are :

- Data warehousing
- Online analytical processing(OLAP)
- Data mining systems

3. **Collaborative CRM :** The collaborative CRM allows the company to synchronize and manage efficient and productive interaction across all the communication channels with :

- Customers
- Prospects

- Partners
- Internal associates

In collaborative CRM, the viewpoint of the customer is taken into account at every transaction level. This helps in providing better services to the customer. Collaborative CRM helps in :

- Reducing web service costs with the help of web collaboration
- Improve the quality of customer service
- Increase customer loyalty

The example of collaborative CRM is the various departments of the company like sales, technical support, marketing etc.

Open Source Customer Relationship Management : There are many versions of CRM that are available in the market. With the help of open source, the programmers can read, redistribute and modify existing source codes to form new software. Some of the widely used open sources CRMs are discussed below :

1. **Sugar CRM :** Sugar CRM is a commercial open source CRM that is suitable for companies of all sizes. The version 3.0 is known as sugar suite. The various functions of this version of sugar CRM are :
   - **Sales Forecasting :** With the help of sales forecasting, the sales manager can calculate potential sales and the associated revenues. It provides a better control over performance, commitment, achieving goals and planning resources.
   - **Campaign and Project Management :** The campaign management tool allows users to perform the following functions :
     - Organize the marketing activities
     - Manage and track marketing efforts on leads, contacts, and opportunities.

     The project management tool enables the users to :
     - Define steps
     - Establish timelines
     - Assign work to members on any project that is useful to the end users.
   - **Email Marketing :** The marketing team performs the following operations :
     - Mail templates
     - Listing of receivers
     - Sending personalized email templates as per the schedule.
   - **Real Time Tracking :** The activities for real time tracking includes :
     - Calendaring
     - Call scheduling
     - Planning meetings
     - Note tracking
   - **Document Management :** The documenting management includes capabilities like :
     - Managing sales documentation
     - Customer reports
   - **Employee Directory :** The customers can be searched on the basis of :
     - First name
     - Last name
     - Roles

– Department
– Contact information

2. **Vtiger CRM :** Vtiger CRM is the most suitable CRM for small and medium business organizations. It was originally built from the source code of Sugar CRM. This CRM is based on open source technologies like Linux, Apache, MySQL and PHP. The features are :
   - **Lead Management :** The lead management creates leads and then converts them to suitable opportunities. It imports important leads from Web, seminars and mails.
   - **Opportunity and Contact Management :** In this, the user can track all the sales opportunities. These opportunities are then associated with accounts, activities, contacts and other nodules. The hierarchy of contacts is maintained so that the customers can be dealt in a proper way.
   - **Account Management** : The user can perform the following operations :
     – Import accounts
     – Export accounts
     – Specify parent-child relationship between accounts
     – Track all the accounts
     – Attach customer specific documents for quick reference
   - **Activity and Inventory Management :** With the help of activity and inventory management, the user can perform the following functions :
     – Add important customer related emails
     – Store all the details of customer meetings and calls
   - **Security Management :** The security management feature helps the organization in defining the role of team members and giving them access as per their role.
3. **Daffodil CRM :** Daffodil CRM is a commercial CRM. It helps to integrate all the aspects of customer life cycle. It helps in automating everyday tasks. The features are :
   - **Printable reports and Export facility :** The daffodil CRM has facilities of :
     – Export to excel
     – Printable report facilities

     The view facility helps to generate different views of data. The users can evaluate the performance of their overall business, sales team and products over a period of time.
   - **Automatic Mail Attachment :** Mails can be attached automatically to the daffodil CRM. Manual selection and sending is not required.
   - **Campaign Facility** : It enables mass mailings to all or selected list of customers. Email templates may or may not be used for this purpose.
   - **Lead Management :** Lead management helps the users in managing and analysing all leads within a single click. A single click provides :
     – All information related to the lead
     – Latest five actions and interactions for follow-up
     – Analysis made by the lead

Customer Relationship Management Areas : There are three areas of the customer relationship management. These are :

1. **Sales :** The sales force predicts and proposes real time analysis of the information that is available and then distributes it to the company. It makes use of the software to

reorganize all the phases of the sales process. Thus it minimizes the time that is required by the sales representatives to spend on each time. The main advantage of this technique is that is allows an organization to use fewer sales representative to manage their clients. In the sales area, there is a sales force automation. The core of the sales force automation is the contact management system. This system is used for tracking and recording each stage in the sales process.

2. **Marketing :** The marketing area concentrates on personalizing the customer prefernces and offer them satisfying experiences. It helps the enterprise to identify and target the possible clients and helps the sales team. The main advantages in this area are :
   - Email search
   - Social media
   - Telephone
   - Direct mails
3. **Service :** The service area is associated with the following :
   - Call centers
   - Email
   - Other communication medias

The call center software helps to direct the customer to the agent that can help them with their current problems. Organizations are considering and using this technology to attract and retain their customers. It helps them to :

- Improve their clients experience
- Increases efficiency
- Minimize costs

Customer relationship management helps in building value driven long term relationships.

Stages in Customer Relationship Management : There are five stages in customer relationship management. These are :

1. Pre-Relationship Stage
2. Early Stage
3. Development Stage
4. Long Term Stage
5. Final Stage.

Models of Customer Relationship Management : There are two models of customer relationship management. These are :

1. **Evans and Luskin Model :** The Evans and Luskin Model came into existence in 1994. The main existence of this model was for effective relationship marketing. The Evans and Luskin Model include :

   (a) **Relationship Marketing Inputs :** These are :
   - Understanding the customer expectations
   - Quality Management
   - Building service partnerships

   (b) **Relationship Marketing Outputs :** These are :
   - Customer satisfaction
   - Customer loyalty
   - Increased profit
   - Quality products

(*c*) **Assessment State :** These are :

- Customer feedback
- Integration

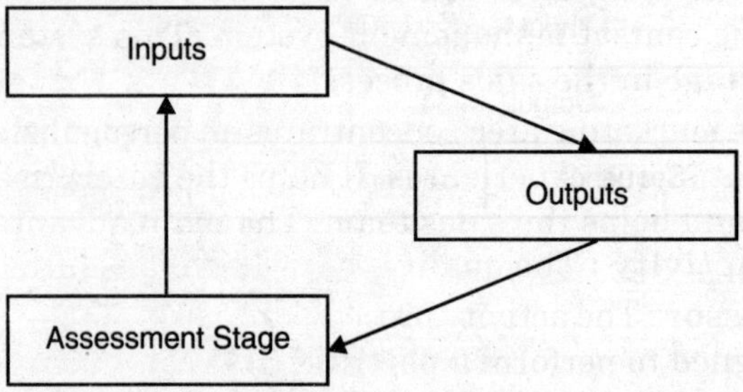

2. **Brock and Barcklay Model :** The Brock and Barcklay model came into existence in 1999. The main existence of this model was for selling partner relationship effectiveness.

Functions of Customer Relationship Management : There are two functions of customer relationship management. These are direct and indirect functions.

1. **Direct Functions :** The direct functions are the basic requirements of a company that are necessary to survive in the competitive market. These are profit, volume and safeguard.
2. **Indirect Functions :** The indirect functions are the actions that are necessary to convince the customer to participate in the various marketing activities. These are innovation, market and access.

Challenges faced by Customer Relationship Management : The challenges faced by a customer relationship management are listed below :

1. Complexity
2. Poor usability
3. Fragmentation
4. Business Reputation
5. Security Concerns

Q.7. How is the defining of subsystem performed?

Ans. The defining of subsystems is done with the help of two main blocks of information. These are :

1. **Conceptual Design :** Conceptual design requires assumptions regarding the subsystems. The conceptual design contains :
   – Review of subsystems
   – Redefining of subsystems if required
   – Detailed activity of each important activity block
2. **Dominant and trade-off performance criterion :** An activity design form is used to develop description about each activity. Trade-off criteria must be used when dominant criteria is not present. The **trade-off criteria** stresses on different functional activities thus aiding in defining the subsystems in a more proper and relevant manner.

<table>
<tr><td colspan="3">Activity</td><td colspan="3">Activity Number</td></tr>
<tr><td colspan="6">Purpose and Description</td></tr>
<tr><td colspan="3">Input</td><td colspan="3">Media</td></tr>
<tr><td colspan="3">Output</td><td colspan="3">Media</td></tr>
<tr><td colspan="2">Sequence</td><td colspan="2">Performer</td><td colspan="2">Decision Rule</td></tr>
</table>

Quality Control Activity : The quality control activity consists of :

- **Activity Processor :** The activity processor consists of the equipment's and the number of workers assigned to perform a particular task.
- **Information Output :** It is the resources available after assigning to a particular activity.
- **Operating Output** : Batches are created. The operating output is the reduced size batches.
- **Information Results :** The information result provides information about the average quality of batches and the number of batches that are inspected in a unit of time.

Activity : The activities can be explained in two terms : Inter-relating activities and hierarchy of activities.

The inter-relating activities must be defined in a network format. The degree of breakdown of the activities shows the size and complexity of the network. The design process will never be completed if the activities are broken down too finely. Important information will be left out from the design process if the activities are broken down too broadly.

The hierarchy of activity network is defined at the functional component level. The subsystems are defined when the activity network is developed. This activity network takes into account each major activity of the conceptual system design.

The subsystem consists of :

- Activities matching to major activity block of conceptual system design.
- Detailed activity blocks : These can be transferred from one group to another so that a network of subsystem can be created.

But now the question arises to how to identify the new subsystem. So, the new revised sub-system can be identified by drawing a loop around the group of activities that are to be included in the subsystem.

Grouping of Activities : The grouping of activities is done on the basis of the following factors :

- Common functions
- Common techniques
- Common inputs
- Common outputs
- Logical flow relationship

Information Components : The information must consist of the following factors. These are :

(*a*) **Available Resources :** The system must be developed in terms of the resources that are available in the company.

(*b*) **Activity :** Each activity and its relationship with another activity must be identified to achieve the system operations and performance.

(*c*) **Information :** The relevant information that is required in the decision making process must be identified. This is done with the help of a computer based system.

(*d*) **Output :** A specific output is required by all the systems. This should include a detailed list of purpose that must be fulfilled by the information output. It should also determine :

- Specific content of each report
- Method of utilization

Information Attaining : There are four sources of obtaining the information. These are :

(*a*) **Task Force Meetings :** The task force meeting consists of the managers and key specialists. It is an important process. The designer must perform the following functions :

- Draw out relevant ideas
- Draw out relevant information
- Abstract ideas
- Include more ideas
- Present the information in a diagrammatic way
- Present the information in documents
- Modify the information
- Evaluate the information

The purpose of the task force meetings is :

- Reduce the information gap
- Increase the operating needs
- Decrease debatable points

(*b*) **Interviews :** The interview process is conducted by the manager and the designer to identify objectives and develop a proper plan. This is performed at both top and intermediate level. It is in addition to the task force meetings. Employees of the organization and specialists should also take part in the interview process.

The designer should search for information on :

- Objectives of the organization
- Format of the information that is presented
- Style of manager on the decision making process
- Resources required for operation of the system
- Resources required for implementation of the system
- Degree of manager's involvement in the decision making process

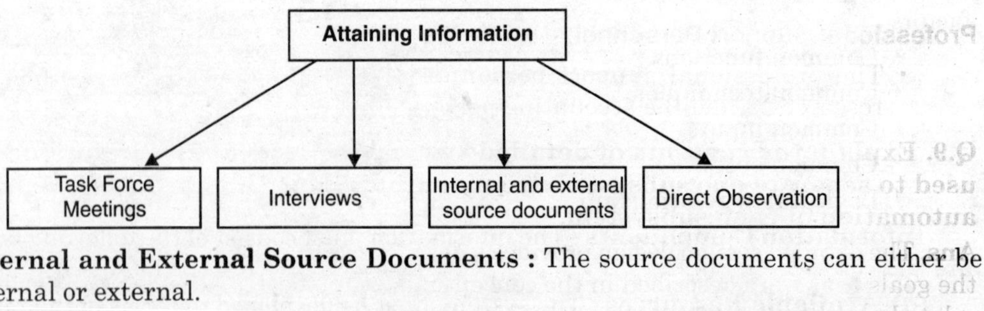

(*c*) **Internal and External Source Documents :** The source documents can either be internal or external.

The **internal** source document provides the system designer with the point of withdrawal. The internal source documents may be large depending upon the size of the company. The traditional reports and documents must be retained.

For Example : Procedure manuals, record of customer complaints

The external document provides :

- Economic information
- Marketing information
- Industrial information
- Financial information

(*d*) Direct Observation : Direct observation also plays an important role in obtaining the information. Designer should also perform on the spot surveys to gather more information and record the relevant data simultaneously. It helps in knowing the physical and environmental factors that are to be accepted in the new system. The designer should organize the data at the end of phase.

Q.8. Why formal training of operating personnel's is mandatory for the success of MIS?

Ans. A program should be developed to impress the management and to train the operating personnel in their new duties.

Role of Management

- The people who participate in the development of the system come under the part of management.
- Here two short seminars are provided. These seminars are enough for these personnel's.
- The first meeting is arranged at the time of detailed designing. Some proposals which are valuable can be combined in the design.
- The second meeting is held at the end of the implementation stage. This meeting usually shows the benefits of the system and the role of the executives.

Train the First Line Supervisors

- First line supervisors should have a proper understanding about the new MIS and its functions.
- They should know about the operations of the system and learn them to operate the system efficiently.

Role of supervisors : The supervisors should have the answers to the following questions :

- The new skills that are to be learned.
- Changes in procedure that are to be made.
- Measurement of performance.
- Jobs will either be upgraded or downgraded.

Professional Support Personnel

- The professional support personnel are computer center personnel, marketing researchers and the accounting personnel.

Q.9. Explain the contents of detailed system design of a MIS. Describe the criteria used to separate one subsystem from another. Explain how to decide the degree of automation of each subsystem.

Ans. The detailed system design provides a detailed explanation of the system that has achieved the goals that were described in the conceptual system design. The detailed system design is related to the design of the operating systems. It is a step by step explanation of the procedure that is required to operate the system. This is the main aim of the detailed system design.

The description in the detailed system design consists of :

- Drawings in the system design
- Flowcharts
- Procedures
- Specification of the equipment's
- Specification of file
- Operating manuals that are required to run the system.
- Operating specifications

The detailed system design cannot be applied on all the systems because of the following reasons :

- Number of different approaches is available for designing the system.
- The design of the system is a rational process.
- The design of the system is a social process.
- The detailed explanation may not give an idea about the general nature of design.
- The designer must interfere frequently with the detailed system design with the help of descriptive essays.

The **dominant criterion** is a criterion that makes an activity important over other activities. The dominant trade off criteria must be identified before undergoing the design steps. For example : Company selling life preserving drugs, company dealing with electric power etc.

The **trade-off criterion** is the criterion in which performance of one activity is reduced to increase the performance of another activity. For example : Criteria of low manufacturing cost are balanced with the public image of the organization that is achieved by reducing the pollution level.

Identification reason of dominant and trade-off criteria : There are two reasons for identifying the dominant and trade-off criteria. These are :

- The decision centers must be identified when the detailed design is developed. It helps to achieve the dominant and trade off criteria.
- A proper management information system must be developed to provide relevant information. It helps in taking decisions regarding the trade-off.
- These personnel's provide input to the Management Information System. They are responsible for processing of the data and Information.
- The concerned personnel's are required to attend one or more orientation meetings.
- These people works with only a small part of the Management Information System. The seminars should be designed in such a way that they provide the personnel's with a complete understanding of the system.

The management can automate the processes to different degrees according to its wish and needs. The degree of automation can be classified as :

(*a*) No automation : Manual processes are used by the people to perform the functions.

(*b*) Work automated, manual control : In this, the work is done by the electromechanical devices but the control is performed manually.

(c) Work automated, manual control, automated feedback : The work is done by the electromechanical devices but the control is performed manually. There is a feedback loop that is controlled by the devices and gives information to the operator.

(*d*) Work automated, automated control, automated feedback : In this no human control is required.

Low level automation is preferred in the following cases :

- Problems are not well structured
- Decision criteria is not well defined
- Gaps in data that is entered into the system
- Cost of human labor is low
- Negligible storage
- Few processing steps
- Processing steps are simple
- Data entered into the system is inconsistent

□□□

PREVIOUS YEAR QUESTION PAPER-2

Q.1. Attempt all the questions in brief :

(*a*) State the management activities.

Ans. Managerial, financial, commercial, accounts, technical and security are the management activities.

(*b*) What are the operations performed by a business organization subsystem?

Ans. The operations performed by the business organization subsystem are financial, design, production and marketing.

(*c*) Explain computer programming.

Ans. The computer programming is concerned with the method to instruct the processor on performing the operations on the input data so that output is produced in the required format.

It consists of different programming languages like C, C++ and Pascal.

(*d*) Explain DBMS engine.

Ans. DBMS engine accepts requests from other database systems and converts them into physical parts. It accesses the database and data dictionary.

(*e*) Explain relational structure.

Ans. Relational structure is a database model that makes use of two dimensional rows and columns to store the data. It is a complex method and is not suitable for the end user. The relational structure is used by the mainframe, midrange and microcomputer systems.

(*f*) Explain program flowcharts.

Ans. Program flowcharts are created by the people who provide instructions to the computer. These flowcharts are the fundamental tool of the programming design. It is a logical sequence of steps that must be done by the computer.

(*g*) What are the commercial applications of ERP?

Ans. Manufacturing, Supply chain management, Financials, Project management and Customer Relationship Management are the commercial applications of ERP.

(*h*) What is daffodil CRM?

Ans. Daffodil CRM is a commercial CRM. It helps to integrate all the aspects of customer life cycle. It helps in automating everyday tasks.

Q.2. Define MIS in various contexts.

Ans. A management information system is designed by an organization so that it can be used by the managers in performing the following functions :

- Directing
- Planning
- Coordinating
- Communicating
- Decision making

The management information system is used by the top management and consists of various controls. This control covers :

- People
- Technology
- Policies

• Procedures

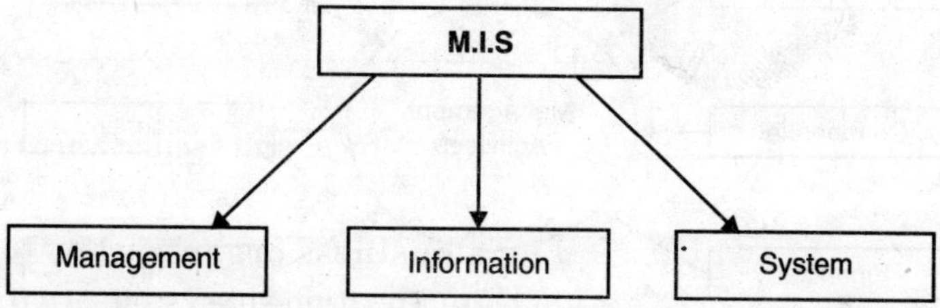

Management : Management is the art of getting the activities done through the people and with the people. It is normally performed by the managers in an organization.

Operations performed by manager : The operations performed by the manager in an organization are :

- **Plan** by making strategies and goals : Managers plan the work by setting strategies and goals and select the best action to complete the task.

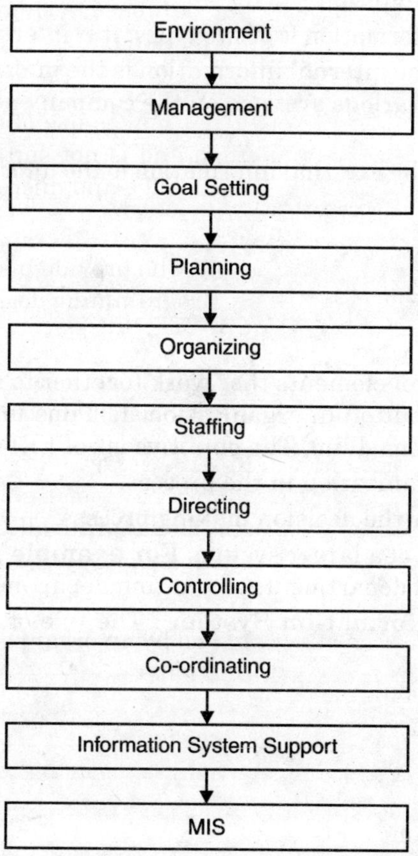

- **Organize** the tasks : Managers organize the tasks that are required to complete the work and group these tasks homogenously.

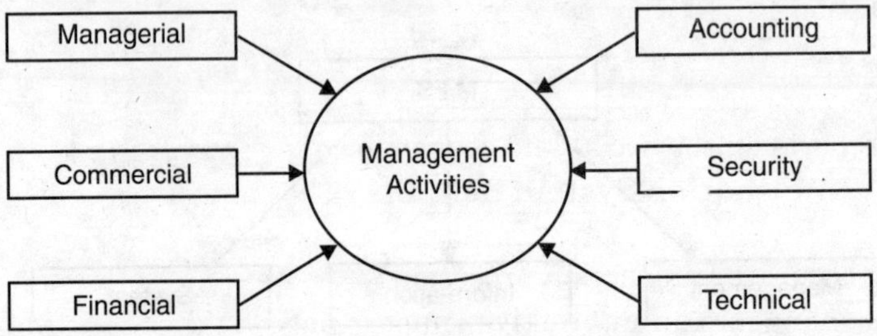

- **Control** the performance : The performance is controlled by the manager by setting some standards and it must be noted that the standards should not be deviated.

Characteristics of Information : The characteristics of information are :

- Updates knowledge level
- Reduces uncertainty
- Helps in decision making
- Improves entity representation

Types of Information : Information is of two types. It is internal and external information.

Internal Information : The internal information is the information that can be gathered easily. It is gathered from the various systems on the company network. For example : Sales figure.

External Information : The external information is the information that can be gathered by :

- Reading articles
- Economic factors
- Social factors
- Financial forecast

System : A system is a set of elements that work together to achieve a common objective. It helps in optimizing the output of an organization. It helps in connecting the subsystems with the information exchange medium. The objective is :

- Provide synergistic organization in the process.
- Provide information for the decision making process.

These subsystems are part of a larger system. **For example :** Organization is the system and departments like financial department, accounting department etc. are its subsystems.

Role of Management Information System : The role of a management information system is :

- Hardware administration
- Software development
- Software support
- End user support
- Database administration
- Corporate data access

Also, management information system plays an important role in providing managers with information like :

- Number of staff required in each department
- Training requirements
- Career development plans

- Job descriptions
- Budget
- Forecasts
- Progress reports on the social and economic conditions

Management Information System Benefits : The benefits of management information system are :

- It helps in accurate transaction processing.
- It provides managers an interactive support in the decision making process.
- It provides organization an edge over other organizations not using MIS.
- Provides better communication.
- Makes better use of resources.
- Makes better use of time by providing facilities every time.
- Improves customer services and relationships.
- Helps in recording and processing all the business dealings and transactions.
- Provides easy access of information to managers and staff.
- Provides relevant information on the past, present and future events that occurred inside and outside an organization.
- Serves as a system framework for organizing business computer applications.

Impact of Management Information System : Management information system creates an impact on the :

- Function
- Performance
- Productivity

of an organization. MIS also creates an impact on the management, managerial efficiency and businesses.

The impacts are :

- Target tracking becomes quite easy.
- Various trends in business processes can be tracked easily.
- It brings clarity in the communication process.
- Creates a sense of professionalism in the business processes.
- Helps in the decision making process.
- Helps the manager to use different tools easily.

Q.3. What is problem definition? Why is it required? Take suitable examples of system and determine its objective and system constraint.

Ans. The first appropriate step that must be taken by the management of an organization is the defining of the problems. These are the problems that are to be solved by an organization. It must be noted that the objectives of an organization are derived with the help of mission statement for a business organization. From the objectives, plans are derived. With each objective and plan, information needs for an organization is derived.

The process of defining the problems is :

(*a*) State the information needs of an organization.
(*b*) Ask questions related to those stated needs.
(*c*) Suggest explanations of that need.
(*d*) Detail the original statement.
(*e*) Review the detailed original statement.

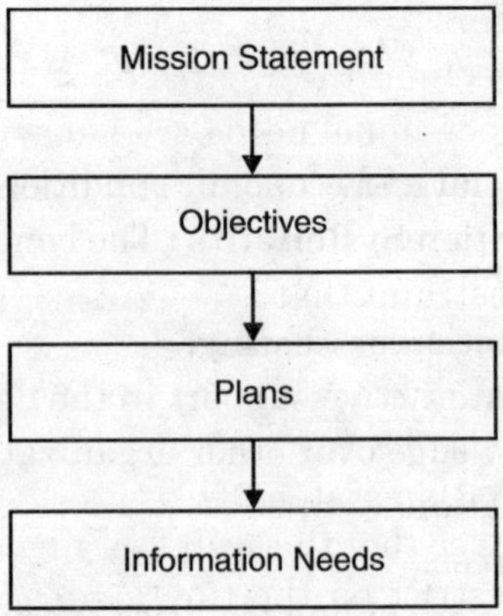

The system objectives of an organization must be set. System objectives are expressed in terms of what the manager can do when the information requirements of an organization is fulfilled.

- The setting of system objectives of an organization that cut across all the functional areas is very difficult.
- It is easy to determine the objectives of an operational system that is served by a main system.

The system objectives are viewed by the following people :

(*a*) **Technician :** The technician sees the system objectives of the smaller system as an input to a larger system.

(*b*) **Manager :** It is the responsibility of the manager to define the system objectives in terms of the information needs. They should not define the terms according to the satisfaction of demands.

The system objectives must be stated in terms of :

- Objectives of the department
- Objectives of the group
- Objectives of the function
- Manager to be served

Establish System Constraints : A constraint helps the designer to modify the conditions under which the objectives can be fulfilled. It also helps the designer to consider the limitations that restrict the design. A constraint can have both positive and negative effect.

Classification of Constraints : Constraints can be classified as internal and external constraints.

Internal Constraints : The internal constraints are of five types. These are :

(*a*) **Top Management Support :** If the top management support is not obtained, then the design cannot be implemented. It is the responsibility of the top management to approve and support the design.

This internal constraint influences on the decision of the management on the kind of the system that is supposed to be designed.

(*b*) **Policy Considerations :** Policy considerations should be kept in mind. It helps in setting the limit for objectives. It also helps in modifying an already existing approach to design a system. The flow of information is influenced by the :

- Structure of the organization

- Position of the managers

The organizational policies limit the approach to system design.

(c) **Personnel Needs :** The personnel needs are a major limiting factor in the :
- Design of Information System
- Utilization of Information System

(*d*) **Cost :** Cost is another constraint that is a resource limitation. It should be noted that the cost to achieve the objectives should be compared with the benefits that will be achieved.

(*e*) **Restrictions :** Restriction is the constraint that is made by the managers and designers on the design of an information system. In the designing phase, the manager needs to alter some of the objectives so that the system is fit with other outputs and equipment. The manager will also restrict on the time and effort that is applied to the entire phase.

External Constraints : The external constraints are of three types. These are :

(*a*) **Customer :** Some of the systems like :
- Order entry
- Order billing

Should be designed keeping in mind the needs of the customer. If some outputs from the system are not acceptable, customer constraint may be faced.

(*b*) **Government :** The governmental constraint is of three types :
- Federal
- State
- Local

This type of constraint imposes restrictions on :
- Maintenance of the security of information
- Meeting of certain procedures of record keeping
- Reporting to stockholders

(*c*) **Suppliers :** The suppliers must also be taken into account during the designing process of an information system.

Q.4. (*a*) What is early system testing and why is it required?

Ans. The early system testing consists of three parts i.e. modeling, simulation and test planning.

(*a*) **Modeling :** Modeling method is used to define the systems in a more proper manner. The modeling method is used for the following functions :
- Determine the quantitative range of input
- Determine the quantitative range of output
- Establish a relationship for performing the transfer functions
- Timely response for the operations in the system

The modeling process consists of **decision tables**. The decision table consists of :
- Quantitative basis
- Qualitative basis

for the decision making process. The decision table is required for both design and implementation of the system.

Format of decision tables : The structure of the decision tables is :
- if these conditions exist
- then perform these actions

(b) Simulation : Testing by simulation is carried out on the :

- System
- Subsystem
- Functional components

Small systems : The testing functions in the smaller organizations are carried out by the conversion to online operations.

Large systems : The testing function in large organizations is quite complex and costly.

Simulation Procedure : The procedure for simulation is carried out in the following steps :

1. Select value of exogenous data like interest rate, cost of raw materials randomly.
2. Input that is non-quantitative in nature is handled by listing all the alternatives and selecting a random value.
3. Trace the effect of exogenous data input throughout the system.
4. Examine the output of various subsystems.
5. Repeat steps 1 to 3 different times.

Activity Process : The activity process can either be automatic or manual.

1. **Manual Activity Process :** In the manual automatic process, errors may occur.
2. **Automatic Activity Process :** The automatic activity process is direct. For example : Computerized systems, decision tables etc.

(c) Test Planning : Individuals show the detailed plans for verifying the final version of the management information system. In the test planning, the following mentioned questions should be answered in a detailed manner :

- What are the user interfaces?
- Which subsystem will be available on different dates?
- Who are the users?
- Who are the managers?

Q.4. (b) Discuss the objectives of detailed system design. What are the important features of project management of detailed system design?

Ans. The detailed system design provides a detailed explanation of the system that has achieved the goals that were described in the conceptual system design. The detailed system design is related to the design of the operating systems. It is a step by step explanation of the procedure that is required to operate the system. This is the main aim of the detailed system design.

The description in the detailed system design consists of :

- Drawings in the system design
- Flowcharts
- Procedures
- Specification of the equipment's
- Specification of file
- Operating manuals that are required to run the system.
- Operating specifications

The detailed system design cannot be applied on all the systems because of the following reasons :

- Number of different approaches is available for designing the system.
- The design of the system is a rational process.

- The design of the system is a social process.
- The detailed explanation may not give an idea about the general nature of design.
- The designer must interfere frequently with the detailed system design with the help of descriptive essays.

Project management of MIS Detailed System Design

Planning and organizing : The first step that is performed in the detailed system design is planning and organizing.

The project management is done for both small and large projects.

- **Small projects :** All the phases must be planned before performing the conceptual system design.
- **Large projects :** All the phases must be planned after the conceptual system design.

The project planning that is done in advance goes wasted, if it is shown by the conceptual system design that the system design is not appropriate.

Amount of effort : The amount of effort that is required in the project management of the detailed system design depends upon :

- Size of the MIS project
- Cost

Project Planning : The project planning of the detailed system design consists of :

(*a*) Establish the project objectives : The establishment of project objectives consists of :
  - Review of the objectives
  - Subdivision of the objectives
  - Modification of the objectives

  These are the objectives that were formed in the conceptual system design.

(*b*) Define the project tasks : Here, the tasks that are to be performed are identified. These tasks are also documented.

(*c*) Plan : Planning of sequential and concurrent tasks is performed. A network diagram showing the events and activities is required.

(*d*) Schedule : As per the requirements of the management, scheduling should be performed. End date and network constraints should be established.

(*e*) PERT diagram : The work and schedule are tied together with the help of PERT diagram.

(*f*) Estimate the labor, equipment and cost required in the project.

(*g*) Budget : A budget should be made for the project. Funds should be allocated to each task.

(*h*) Plan the staffing organization.

Project Control : In the controlling of project, following factors are taken into consideration.

(*a*) During the progress of the project, determine whether the project objectives are fulfilled or not.

(*b*) The expenditure of the project must be evaluated in terms of :
  - Time
  - Work done

  The budget must be revised if required.

(*c*) Evaluate the work force utilization and work performed by an individual.

(*d*) Evaluate time, cost and work requirements in terms of :
- Schedule
- Budget

Q.5. What are the steps in planning implementation of MIS?

Ans. Implementation is an important managerial activity related to the adoption and management of an innovation. It is an activity which starts when the system is developed and is ready to be installed in the organization.

The design of the Management Information system seems to the management as an expensive project. The implementation of MIS is done when the output of the MIS are continuously used by the decision makers. So basically implementation is a process of introducing a newly developed Management Information System into an organization for the use of end user, owners etc.

Steps before the implementation process occur : The appropriate steps that must be taken before the implementation process is started are :
- Designer must make an appropriate environment so that he gets full support from all the levels of management in an organization.
- The user requires adequate training so that he can learn to implement the system efficiently and effectively.

The implementation outcome can be subjected to various factors like :
- Role of users in the implementation process.
- Amount of management support for the implementation process.
- Complexity of risk involved in the implementation activity.
- Level of risk involved in the implementation activity.
- Quality of management in the implementation process.

Plan the Implementation : Planning the implementation of MIS in an organization is the first step in the implementation phase. In planning the implementation of the MIS, there are three main phases which occurs in a series. These three phases are listed below :
- Initial Installation
- Test of the whole system
- Evaluation, maintenance and control of the system

The planning process is a function of line management. The systems analyst must prepare plans to assist the managers but the managers are involved in last, only if required. Also,the manager requires the services of the system analyst to make detailed plans. The manager prefers to make decisions by using the most recent information. This recent information includes :
- Management Information System Specification
- Proposed plans of the System Analyst
- Current operating situation

The planning phase of the implementation process must include the following :
- Conclusion of the design process
- Documentation of the system

Many implementation activities should be started in parallel to reduce the implementation time.

For Example : Training of personnel and preparation of software can be done in parallel with each other and with other implementation activities.

(a) Identify the Implementation Tasks : The major implementation tasks consists of :

- Planning the implementation activities
- Acquiring and laying out facilities and offices
- Organizing the personnel for implementation
- Developing procedures for installation and testing
- Developing the training program for operating personnel
- Completing the system's software
- Acquiring required hardware
- Generating files
- Designing forms
- Testing the entire system
- Completing cutover to the new system
- Documenting the system
- Evaluating the MIS
- Providing system maintenance (debugging and improving)

(b) Establish Relationships among Tasks : The establishment of relationship among tasks is done for both small and large projects.

- **Small Projects :** For the small projects, the order of performance may be described in a text format. A Gantt chart or network diagram makes visualization of the plan and schedule much clearer.
- **Large Projects** : For large projects, many parallel and sequential activities are interrelated so that a network diagram can also be used.

(c) Establishing of MIS : Management Information System should be established because of the following reasons :

- Monitoring the progress of Implementation
- Proper control of activities
- Efficient Information system should be developed.

(d) Establish a Schedule : Schedule is prepared by the system designers. The role of the system designer is to :

- Estimate the time between the events in a program network.
- The critical path that is the longest time through the network can also be calculated.
- End date is established by knowing the starting date.

(e) Prepare a Cost Schedule tied to tasks and time : The cost of completing each task is established as part of the plan. The rate of expenditure should also be calculated.

(f) Establishing a Reporting and Control system : Establishing of a Reporting and Control system is performed by the key people involved :

- Reporting of the work in progress should be obtained in weekly meetings.
- Control of the work in progress may be obtained by weekly meetings.
- Financial personnel must make sure that report formats allow them to show cost and the technical progress relationship.

The report establishes two relationships :

- Cost and time relationship
- Cost and technical progress relationship

The objective of control system is to minimize the confusion, delay and costs.

Q.6. Explain procurement management system.

Ans. A procurement management process or a procurement process is the method of purchasing goods and services from external suppliers. Obtaining goods and services from external suppliers can be a critical path for many projects. The performance of the supplier reflects the performance of the overall project team. The supplier's performance should be monitored carefully and regularly so that the delivery of the product meets the expectations of the team.

The goods and services should be obtained at the best possible cost to meet the needs of the purchaser in terms of :

- Quality
- Quantity
- Time
- Location

The management process provides procurement procedures. This procedure consists of step-by-step explanation of how to purchase goods from suppliers.

Thus the procurement management process helps you : "Get What You Have Paid For".

Procurement Management Functions : The procurement management process consists of the following functions :

- Identify the goods and services that are to be procured
- Managing the orders
- Managing the receipts
- Review of items from the suppliers
- Approval of items from suppliers
- Managing of the relationship with the supplier
- Ensure that a high level of service is received
- Review the supplier performance against the contract
- Identify and resolve supplier performance issues
- Approve the supplier payments
- Communicate the status to the management

Types of Procurement Management : The classification of the procurement management is based on the basis of the consumption of goods. There are two types of procurement management. These are direct and indirect procurement.

Direct Procurement : The direct procurement is production related. It takes place during the manufacturing process only. It consists of the items that are part of the finished product. For example : raw material, components and parts. It affects the production process of the manufacturing firms.

Features of Direct Procurement : The features of direct procurement are :

- Large quantity
- High frequency
- Industry specific value
- Operational in nature

For Example : Crude oil in petroleum industry.

Indirect Procurement : The indirect procurement is non-production related. It deals with the operating resources. Operating resources are the resources that are purchased by a company to enable its operations.

Features of Indirect Procurement : The features of indirect procurement are :

- Low quantity
- High frequency
- Low value
- Tactical in nature

For Example : Lubricants and spare parts.

Procurement Management Life Cycle : The procurement management life cycle consists of eight steps. These are :

1. **Information Gathering :** It is the responsibility of the sales and marketing team to gather the appropriate suppliers who can fulfil specific requirements.
2. **Supplier Contract :** When the supplier has been identified, following operations must be performed :

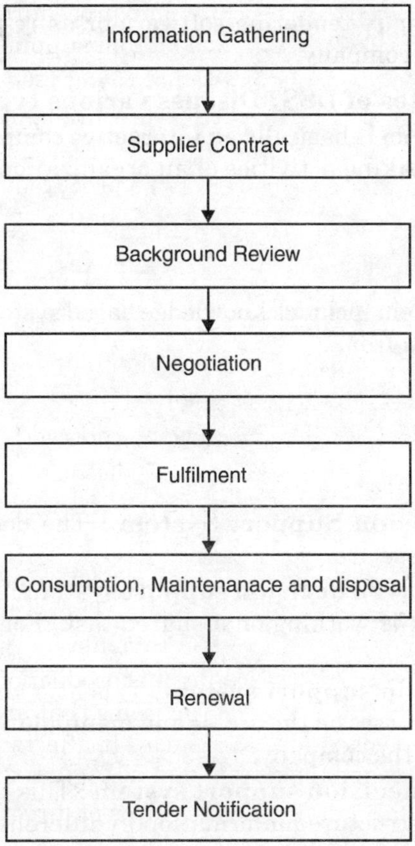

   - Request for Quotation
   - Request for Proposal
   - Request for Information
   - Request for Tender
3. **Background Review :** In the background review, reference for product and service quality is consulted. Requirements for services like installation, maintenance and warranty is investigated.
4. **Negotiation :** Negotiation for price and delivery schedule is done.

5. **Fulfilment :** The following operations are done :
   - Shipment delivery
   - Payment
   - Installation
   - Training
6. **Consumption, maintenance and disposal :** In this phase, the company evaluates the performance of the services.
7. **Renewal :** When the services are consumed, the contract expires. If it has to be reordered, the company must determine whether to continue with the sample supplier or some another supplier.
8. **Tender Notification :** Many institutions make use of the notification service in order to raise the competition for the chosen opportunity. These systems can either be direct with the help of their re-tendering software or as re-packaged notification from an external notification company.

Q.7. What are the attributes of DSS? Discuss various types of DSS?

Ans. A decision support system is basically an interactive computer based information system that supports the decision making activities of an organization. It serves the :

- Management level
- Operations level
- Planning level

The decision support system includes knowledge based system. It helps the decision makers to compile useful information from :

- Raw data
- Documents
- Business models
- Personal knowledge

Classification of Decision Support System : The decision support system can be classified into five parts. These are :

(*a*) **Communication-driven decision support system :** This type of DSS supports more than one person that is working on a shared task. For example : Groove, Microsoft's NetMeeting.

(*b*) **Data-driven decision support system :** It is also known as data oriented decision support system. It stresses on the access and manipulation of the time series of internal and external data of the company.

(*c*) **Document-driven decision support system :** This type of DSS manages, retrieves and manipulates unstructured information in different electronic formats.

(*d*) **Knowledge-driven decision support system :** This DSS provides specialized expertise in problem solving. It is stored as facts, rules and procedures.

(*e*) **Model-driven decision support system :** This type of DSS makes use of data and parameters that are provided by the users to help in the decision making process.

Components of Decision Support System : There are three fundamental components of decision support system. These are :

- **Database :** The database is also known as knowledge base.
- **Model :** The model is for either decision context or the user criteria.
- **User interface :** Users are an important part of the DSS architecture.

Decision Support System Framework : The framework of decision support system includes :

- People
- Technology
- Development approach

Components : Decision support system consists of :

- **Application :** This must be the actual application that can be used by the user. It helps the decision maker to make decisions about a particular problem.
- **Generator** : The generator consists of an environment that allows people to develop DSS specific applications. Case tools are used by the users. It also consists of special languages, linking modules and function libraries.
- **Tools :** The tools can be lower level hardware or software.

The decision support system consists of :

- **Inputs :** Input can be factors, numbers and characters that can be analysed.
- **Expertise** : Inputs require manual analysis by the user.
- **Outputs** : The output is the transformed data. It helps to generate decision support system decisions.
- **Decisions** : Decisions are the results that are generated by the decision support system. It is created on the basis of user criteria.

Decision Support System Applications : The applications of decision support system are :

- It is used for medical diagnosis.
- It is used in business and management.
- It is used in marketing for sustainable development.
- It is used in forest management.

Decision Support System Benefits : The benefits of decision support system are :

- It improves personal efficiency.
- It helps in speeding up the process of decision making.
- It increases organizational control.
- It increases the speed of problem solving in an organization.
- It promotes learning.
- It promotes training.
- It helps in automating the managerial processes.
- It creates a competitive advantage over competition.

Q.8. Write short notes on :

(*a*) Detailed system documentation.

Ans. Documents' specifying the system, operation and design is the end product of the detailed system design. The documentation of the detailed system design must consist of :

- Summary flowchart
- Detailed flowchart
- Specification of database
- Hardware requirements
- Software requirements
- Performance specifications

- Cost of operation

The documentation should be in the following format :

- Standardized format
- Input and output activity must be linked
- Good reporting style should be used

(b) Role of manager in MIS.

Ans. It is the responsibility of the manager to define the system objectives in terms of the information needs. They should not define the terms according to the satisfaction of demands.

The manager's style affects the kind of information that is required. The style includes :

- Technical background of the manager
- Leadership quality of the manager
- Decision making ability of the manager

The view of the information system is different for different managers. Most of the managers do not pay attention on the type of information that is required.

Q.9. Write notes on :

(a) RDBMS.

Ans. A relational database management system is a database management system that is based on a relational model. It is designed by E.F. Codd. It is based on the theory of sets and relations. It is the most popular database.

- Relational model represents data in a table. The table contains an array of rows and columns.
- The row contains data related to an entity.
- The column contains data related to single attribute of the entity.

Advantages

- It is simple in nature.
- It is easy to understand.
- It is easy to manipulate.
- It allows having a dynamic relationship between the entities.

(b) Components of a Computer system :

Ans. The components of a computer system are :

A. **Input :** Input is the data that is required to enter into the computer system. The input data should be in machine readable format. The data that is received from punch cards, magnetic tapes, keyboards are converted into machine readable format.

B. **Central Processor :** The central processor is the main part of the computer. It consists of three parts. These are :

- **Control Section :** The control section is responsible for integrating all the operations of the computer system. It basically directs and coordinates the operations.
- **Control input-output devices :** The control of input and output devices is associated with two factors. First is the entry and removal of the information from the storage devices. Second is the routing of information between the storage devices and arithmetic logic section.
- **Arithmetic logic section :** The arithmetic logic section performs the arithmetic and logic operations. The arithmetic operation is the calculation process and shifting

of numbers. The logic operation is the decision to change the sequence of execution of instructions.

C. **Storage :** The data should be placed in the storage devices before performing the processing functions on the data. Storage is of two type's i.e. internal and external storage.

Internal storage : Internal storage is known as memory. It allows the machine to store data in electronic form. It allows the machine to store long instructions that is required to guide the machines regarding what to do.

External storage : External storage is further classified into two types. Direct access and sequential are the two types of external storage.

- **Direct access :** In the direct access, the data can be read from anywhere. It is not necessary that the required data should be read from the beginning of the file. For Example : disk and magnetic drums.
- **Sequential :** The sequential storage requires that the data must be read from the beginning of the file so that the desired record is read and written. For Example : magnetic tapes.

D. **Output :** The output is the final result of the data processing. The output is recorded on different media's like cards and magnetic media. The record information can either be updated.

Q.9. (*c*) Computer Hardware.

Ans. The **hardware** is the physical parts of the computer. The computer hardware can be classified as :

(*a*) **Motherboard :** The motherboard of the computer is the central circuit board that makes up the computer system. It holds up :
- CPU
- Main memory
- Slots for expansion cards

(*b*) **Storage :** Storage of a computer system means those parts of the computer system that stores data for some period of time.

(*c*) **Input-Output devices :** The input-output device refers to the various interfaces that different functional units of system use to communicate with each other.

(*d*) **Power supply :** It is a case that holds :
- Transformer
- Voltage control
- Fan

(*e*) **Interface controller :** Interface controller like USB, firewire is used to connect the computer to external peripheral devices like printers, scanners etc.

□□□

PREVIOUS YEAR QUESTION PAPER-3

Q.1. Attempt all the questions in brief :

(a) What are the phases of procurement management lifecycle?

Ans. Information gathering, supplier contract, background review, negotiation, fulfilment, consumption, maintenance and disposal, renewal and tender notification are the phases of procurement management lifecycle.

(b) What are channel management tools?

Ans. Channel management tools aims to understand how the customer interacts with the company. The aim of the channel management tool is to deliver the products and services across multiple channels in an effective, efficient and consistent manner.

(c) What are the three integration strategies?

Ans. Push, pull and push | pull are the three types of integration strategies.

(d) Explain the pull strategy.

Ans. In the pull strategy, both the production and distribution is demand driven. It is a fast information flow mechanism. The pull strategy makes efficient use of resources. Dell and Amazon follows the pull strategy.

(e) Explain the globalization era.

Ans. The characteristic of the globalization era is the attention provided to the global systems of supplier relationships and the expansion of supply chains over national boundaries. The characteristics of the globalization era are competitive advantage, value adding and reducing costs through global sourcing.

(f) What are the items that must be documented?

Ans. Plans, project and system objective, specification of functions, specification of performance, user interface specifications, user instruction manual, reference manual and maintenance guideline are the items that must be documented.

(g) What are the components of an ERP?

Ans. Transactional Backbone, advanced application and management portal are the components of an ERP.

(h) Explain direct approach.

Ans. In the direct approach, the new system is installed directly with the immediate discontinuance of the old existing system. It is also known as immediate cut over or change over.

Q.2. How information needs are determined? How the information is collected?

Ans. The information needs are very necessary in designing the information system. There should be a clear statement of the information need. Thus the information need is the fundamental and necessary requirement in designing an information system.

Nowadays, many companies spend large amount of money on the hardware and software of the system without determining the actual needs of an information system. It is thus the responsibility of the manager to determine the information needs otherwise the information system will produce fewer results.

The determination of the information needs includes :

1. Personal Attribute
2. Organizational Environment

(a) Information System Knowledge : If the managers are aware about the working and operations of the information system, it helps in making more appropriate information requests.

(b) Managerial Style : The manager's style affects the kind of information that is required. The style includes :

- Technical background of the manager
- Leadership quality of the manager
- Decision making ability of the manager

(c) Managers view of the Information needs : The view of the information system is different for different managers. Most of the managers do not pay attention on the type of information that is required.

Organizational Environment : The organizational environment includes :

(a) Nature of the Company : It is the nature of the company that decides functions like :

- Communication
- Operations

These functions depend upon the company's size and complexity. It must be noted that the larger the firm is more formal information systems are required.

(b) Management Level : The management level includes three levels. These are :

- **Strategic Planning :** The strategic planning includes :
  - One time report
  - Inquiry
  - Summary
- **Management Control :** The management control includes :
  - Exception report
  - Different regular reports
  - Summary
- **Operational Control :** The operational control includes :
  - Formal reports
  - Fixed procedures
  - Daily transaction report

(c) Organizational Structure : The information needs can be easily determined with the help of organizational structure. The more highly structured the organization is, the easier is to determine the information needs.

The information needs of the managers can be easily determined if the following factors are always considered :

- Understanding the relationships
- Defining decision making areas properly

Determine Information Sources : Determination of the information needs and information sources are quite corresponding. The natural way of finding information is through :

- Books
- Records
- Files

The determination of information sources consists of two theories :

(a) Redesigning of the new system from the old system : If a small modification is made in the existing system, then the information requirements can be fulfilled. This helps in reducing the designing effort. A study should be done to know the specific areas that need improvement. A study should be done for knowing the :

- Common input sources
- Data linked with the new design
- Cost associated with the new design

(*b*) **Logical Approach :** The logical approach of determining the information sources is also known as fresh approach. The logical approach says that the detailed analysis of the existing system is not required. This is not necessary because :

- New system will be changed.
- New system should not be predicted on the basis of the old one.

Choice of Approach : Now the question arises, which approach should be used? The choice of approach is done :

- It is done by the manager or designer.
- It is done on the basis of the information sources.
- It is done on the basis of the objectives of the system that is being designed.
- It is done on the preference of the old or new system.

Analysis and integration of the Information Sources : The source of information is broadly classified into :

(*a*) **Internal and External Records :** The records are of two types. One is internal and another is external record.

Internal Record : Internal records are of written type. These documents contain information about the existing system. For example : file record, letters, reports and documents.

External Record : External record is the record that comes from various different sources. For example : Government Statistics.

(*b*) **Interview Process :** Interview process of managers and personnel's is a method of identifying the possible sources of information. It helps in analyzing the existing information system. The interview process helps in obtaining secure information.

The **problem** faced while conducting interviews is human made errors. The **solution** to overcome this error is through :

- Proper planning
- Increasing confidence level

(*c*) **Sampling Method :** Sampling method is performed when there is a large amount of data available. In this method, only a small part of data is examined. The advantage of sampling method is :

- Less time is required to perform the sampling operation.
- Less cost is faced to perform the operations.

For Example : Work Sampling.

The work sampling method is used to perform the following operations :

- Action of people
- Action of machine

These operations are sampled in terms of time.

Analysis and Synthesis : Analysis and synthesis makes use of two techniques to avoid duplication. These techniques are :

(*a*) **Input-Output analysis :** The analysis on input and output is done on the basis of input-output chart, visual display of the information inputs and information output.

(b) Multi-Dimensional Flow : Multi-dimensional flow is a technique of organizing the various information sources. This technique makes use of the flowchart. The flowchart helps in tracing the flow of information from source to destination. The multi-dimensional flow flowchart should be arranged in an increasing order.

Q.3. (a) What do you mean by alternative conceptual design?

Ans. Alternative conceptual designs should be made. We must briefly note the various constituents of the conceptual design. These are :

- Decision points
- Information flow patterns
- Information channel
- Manager's role
- Relationship between the operating and management information system.

The alternative designs should be measured in terms of its advantages and disadvantages.

Evaluation method : The evaluation method is a method to compare the alterative designs. The method consists of the following criterion :

- The objectives of the system must be compared with the expected performance of the design of the system.
- Prepare a cost effective analysis of the system.
- Flowcharts must be examined. It helps to identify the strong and weak points of the system design.
- Study the design in more detail.
- Study the number of operations performed by the system and its potential.
- Examine the database quality.

Q.3. (b) How is acquiring of floor and space done?

Ans. Floor Space Layouts : Installation of a new system to replace the current system requires many facilities like new office, computer rooms and production layouts.

The MIS project managers should perform the following operations :

- Prepare rough layouts of the MIS
- Estimate the floor area required
- Prepare a cost estimate and submit it to the management for approval

Plan Space Layouts : Space planning should be started as soon as the approval is obtained regarding the space allocations from the management. This should be done at an earliest mainly because of two reasons :

- There may be a long lead time if new partitions, electrical work, air conditioning or even newbuildings are required.
- The detailed work flow depends upon the physical arrangements of the buildings.

Space Planning Includes

- Space occupied by the people
- Space occupied by an equipment
- Movement of the people and equipment's in the work process
- Number of exists
- Kind of exists
- Location of utilities, outlets and controls
- Environmental requirement for an equipment

- Safety factors
- Working condition of the personnel

A large investment in good working conditions will repay its cost many times in the future.

Q.4. Explain supply chain management in detail.

Ans. Supply chain management are a set of approaches that are used to efficiently integrate the suppliers, manufacturers, warehouses and distribution centres. It is also the management of a network of interconnected businesses that are involved in the delivery of product and service packages that is required by the end customers.

The approaches are performed so that the product is produced and distributed in :

- Right quantity
- Right location
- Right time

Supply chain management is also known as logistics network.

Supply chain management is the designing, planning, controlling, monitoring of the supply chain activities with an objective of creating net value, measuring performance globally and synchronizing supply with demand.

Inputs of Supply Chain Management : There are three inputs of supply chain management. These are :

- Raw material
- Working in inventories
- Finished products

Supply Chain Elements : There are four supply chain elements. These are strategic, management tactical and operational elements.

1. **Strategic Elements :** The strategic elements are the following :
   - **Strategic network optimization** : It includes the number, location and size of the distribution centres.
   - **Strategic partnerships** : This is the partnership with the suppliers, distributors and customers. It provides the communication channel for sharing critical information.
   - **Product lifecycle management** : It is the integration of the new products with the existing ones into the supply chain activities.
   - Supply chain design
   - Long term planning
   - Resource attainment
2. **Tactical Elements :** The tactical elements are the following :
   - **Production decisions** : These include contracting and scheduling.
   - Distribution planning
   - Production planning
   - Medium term planning
   - Resource allocation
   - Purchasing decisions
   - **Inventory decisions :** These include quantity, location and quality of the inventory.
   - **Transportation strategy :** These include routes and contracting.
   - Customer demands and habits.

3. **Operational Elements :** The operational elements are the following :
   - Shipment scheduling
   - Short term planning
   - Resource scheduling
   - Production scheduling
   - Demand planning and forecasting
   - **Inbound operations :** It includes the transportation form suppliers.
   - **Outbound operations :** It includes warehousing and transportation to the customers.
   - **Production operations :** It includes the consumption of materials and flow of finished goods.
4. **Management Elements :** The management elements are the following :
   - Planning and control
   - Work structure
   - Organization structure
   - Risk and reward structure
   - Culture and attitude
   - Management methods
   - Power and leadership structure

Supply Chain Management Strategies : The supply chain management strategies are of two types. These are operation and integration strategies.

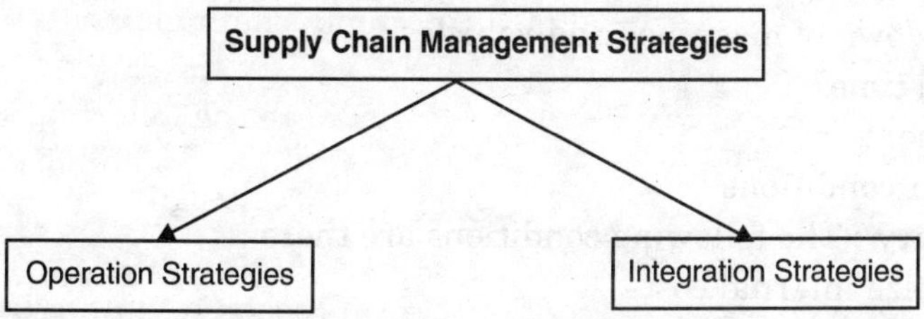

Operation Strategies : The operation strategies are of three types. These are :

- **Make to Stock :** The stock must be chosen when the products are standardized and demand is predictable. The benefits of make to stock are that the manufacturing costs are lower and the demands of the customers are fulfilled quickly.
- **Make to Order :** The order must be chosen when the products are customized. The benefits of make to order are that it improves the security levels and reduces the inventory cost.
- **Configure to Order :** The order must be chosen when there are various variations on the finished products and the demand is infrequent.The benefits of configure to order are the low inventory costs and simple planning.

Integration Strategies : The supply chain integration strategies are of three types. These are :

- **Pull Strategy :** In the pull strategy, both the production and distribution is demand driven. It is a fast information flow mechanism. The pull strategy makes efficient use of resources. **Examples :** Dell and Amazon follows the pull strategy.
- **Push Strategy :** In the push strategy, the manufacturing forecasts are of long range. There is an inefficient usage of the production facilities. The push strategy takes longer time to react to the market changes. **Example** : Auto Industry.

- **Push | Pull Strategy :** The push | pull strategy is the hybrid of push and pull strategy. This strategy overcomes the disadvantage of both the push and pull strategy. The early stage of product assembly is done with the help of **push** strategy. The final staging is done with the help of **pull** strategy. The advantage of using this strategy is that it helps to reduce the uncertainty. The final product assembly done is on the demand of the customer. **Example :** Cisco Systems.

Key Issues in Supply Chain Management : There are five main key issues of a supply chain management. These are :

(*a*) Forecasts are never right

(*b*) Longer the forecast horizon, the worse is the forecast

(*c*) Aggregate forecasts are more accurate

(*d*) Network Planning

(*e*) Supply Contracts

Benefits of Supply Chain Management : The benefits of the supply chain management are :

- Delivery performance
- Inventory reduction
- Lower supply chain costs
- Service level requirements are satisfied.

Difficulties faced by Supply Chain Management : The difficulties faced by the supply chain management are :

1. **Uncertainty :** The following conditions are there :
   - Breakdown of machines and vehicles
   - Travel time
   - Weather
   - Labour conditions
2. **Complexity :** The following conditions are there :
   - Minimize internal costs
   - Minimize uncertainty
3. **Distribution Network Configuration** : The following conditions are there :
   - Number of suppliers
   - Location of suppliers
   - Network of suppliers
   - Production facilities
   - Distribution centres
   - Warehouses
   - Customers
4. **Distribution Strategy** : The following conditions are there :
   - Questions of operating control i.e. centralized, decentralized or shared.
   - Delivery scheme i.e. direct shipment, pool point shipping, direct store delivery, closed loop shipping.
   - Mode of transportation i.e. motor carrier, truckload, parcel, railroad.
   - Replenishment strategy i.e. pull, push or hybrid.
   - Transportation control i.e. private, common and contract carrier.
5. **Information :** The following conditions are there :

- Demand signals
- Forecasts
- Inventory
- Transportation
- Collaboration

6. **Inventory Management** : The following conditions are there :
   - Quantity of inventory
   - Location of inventory
   - Work-in-process
   - Finished goods.
7. **Cash-Flow** :The following conditions are there :
   - Arranging the payment terms
   - Methodologies for exchanging funds across entities within the supply chain.

Applications of Supply Chain Management : The applications of the supply chain management are :

- Globalization
- Increased cross border sourcing
- Collaboration for parts of value chain with low-cost providers
- Shared service centres for logistical and administrative functions
- Increasingly global operations, which require increasingly global coordination and planning.

Q.5. Write short notes on :

(*a*) Decisions support system.

Ans. A decision support system is basically an interactive computer based information system that supports the decision making activities of an organization. It serves the :

- Management level
- Operations level
- Planning level

The decision support system includes knowledge based system. It helps the decision makers to compile useful information from :

- Raw data
- Documents
- Business models
- Personal knowledge

Classification of Decision Support System : The decision support system can be classified into five parts. These are :

(*a*) **Communication-driven decision support system :** This type of DSS supports more than one person that is working on a shared task. For example : Groove, Microsoft's NetMeeting.

(*b*) **Data-driven decision support system :** It is also known as data oriented decision support system. It stresses on the access and manipulation of the time series of internal and external data of the company.

(*c*) **Document-driven decision support system :** This type of DSS manages, retrieves and manipulates unstructured information in different electronic formats.

(*d*) **Knowledge-driven decision support system :** This DSS provides specialized expertise in problem solving. It is stored as facts, rules and procedures.

(*e*) **Model-driven decision support system :** This type of DSS makes use of data and parameters that are provided by the users to help in the decision making process.

Components of Decision Support System : There are three fundamental components of decision support system. These are :

- **Database :** The database is also known as knowledge base.
- **Model :** The model is for either decision context or the user criteria.
- **User interface :** Users are an important part of the DSS architecture.

Decision Support System Framework : The framework of decision support system includes :

- People
- Technology
- Development approach

Components : Decision support system consists of :

- **Application :** This must be the actual application that can be used by the user. It helps the decision maker to make decisions about a particular problem.
- **Generator** : The generator consists of an environment that allows people to develop DSS specific applications. Case tools are used by the users. It also consists of special languages, linking modules and function libraries.
- **Tools :** The tools can be lower level hardware or software.

The decision support system consists of :

- **Inputs :** Input can be factors, numbers and characters that can be analysed.
- **Expertise** : Inputs require manual analysis by the user.
- **Outputs** : The output is the transformed data. It helps to generate decision support system decisions.
- **Decisions** : Decisions are the results that are generated by the decision support system. It is created on the basis of user criteria.

Decision Support System Applications : The applications of decision support system are :

- It is used for medical diagnosis.
- It is used in business and management.
- It is used in marketing for sustainable development.
- It is used in forest management.

Decision Support System Benefits : The benefits of decision support system are :

- It improves personal efficiency.
- It helps in speeding up the process of decision making.
- It increases organizational control.
- It increases the speed of problem solving in an organization.
- It promotes learning.
- It promotes training.
- It helps in automating the managerial processes.
- It creates a competitive advantage over competition.

(b) Soft spots in MIS planning.

Ans. Soft spots in planning are related to the Management Information System that is under planning and development.

Review

- **MIS response to a Business Plan :** The main problems that are faced by the MIS planning are that the personnel's associated with the planning process forgets the original basic purpose of the management. The professionals face difficulties for responding just to the information needs of the main business. One of the difficulties is the temptation to purchase the latest hardware and software techniques. When a change is made to the business plan, the corresponding MIS must be made and changed. Each MIS plan must be a proper response to a business plan.
- **One for one conversion :** The one for one conversion provides an opportunity to upgrade an existing system into a Management Information System. Suppose a technician converts an existing computer system or a manual and converts it without upgrading or changing it.
- **System View :** In a system view, there should be a master plan available with the help of which MIS planning and development can be performed. Without a master plan, a system failure occurs.
- **Setting Project and System Objectives :** System objectives are required for the development of a correct plan. Also, objectives cannot be made until some of the general planning variables are known. Thus both the setting up of planning and objectives goes hand in hand.
- **Facing Constraints :** Constraints are faced on financial definition, system performance, system Cost and development schedule. An MIS problem occurs if freedom from these constraints is obtained. It is important that both the management and the technicians must recognize the importance of the constraints and plan accordingly.
- **Plan to Sale the MIS :** System designers face the problem of gaining acceptance from the user for whom the system is designed. Thus it should be taken care that the system should be deigned in such a way that it is easily accepted by the users. It is the toughest part of the entire designing and implementation phase.
- **Detail Planning :** The MIS developers must plan in detail each phase of the project.

Q.6. What is system approach? What is its relevance?

Ans. System approach is an approach that is basically designed to make use of scientific analysis in complex organizations. It consists of two operations i.e. developing and managing operating systems and designing information systems for decision making process.

Increased Complexity of Business : The increased business complexity consists of four factors :

(a) Technological revolution : The technological revolution consists of :

- Automation techniques are adopted by a large number of industries.
- Industries like transportation, communication, agriculture and manufacturing undergo a vast change in :
  - Products
  - Techniques
  - Outputs
  - Productivity

- These changes occur at a high speed and the management requires a large number of steps to cope with this change.
- The technological revolution requires managerial revolution.

(b) Research and Development : The research and development consists of :

- The technological change is occurring because of the changes in the research and development.
- A large amount of money is being spent upon the research and development.
- Nowadays the products and its operations are becoming quite complex. The lifecycle of the product has become shortened.
- Better planning should be provided.
- Better management should be provided.
- Better information should be provided.

(c) Product changes : The product change consists of :

- Product change has occurred because of the following factors :
  - Advancement in technology
  - Research and development
  - Customer sophistication
- The modern organization has to produce an optimized product in a shorter time.

(d) Information explosion : The information explosion consists of :

- The information explosion has an impact on the :
  - Management
  - Organization
- The modern manager must know that :
  - Ability to obtain the information
  - Store the information
  - Retrieve the information
  - Display the information

  is very important.

Increased Complexity of Management : The increased management complexity consists of four developments :

(a) Information feedback systems : The information feedback system consists of :

- The information feedback system is concerned with the use of information for the purpose of control.
- It is applied to :
  - Business systems
  - Management systems
  - Engineering systems
  - Biological systems
- The output of the system generates an output. It causes an action that corrects the output leading another decision.
- The examples of information feedback systems are :
  - Thermostat-furnace temperature
  - Missile system

(b) Decision making : The decision making consists of :

- The decision making process is a consideration of management and information design.
- The decision making process is based upon :
  - Policy
  - Rule
- These are programmed for certain computer applications. Faster and accurate operations are done for these computer applications.
- The examples of decision making are :
  - Payroll
  - Customer billing
  - Customer purchasing

(c) Management Science : The management science consists of purpose and tools.

- The purpose of management science are :
  - Design of programmed decision rules
  - It helps the managers to take complex decisions.
- Simulation tool is used for the construction of a mathematical model for a system that is under study.
- Effects of different management policies, resources etc. can be predicted before the final decision because behavior of the model under manipulation simulates the behavior of the real system.

(d) Electronic computer : The electronic computer consists of :

- The electronic computer helps in storage of data.
- It helps in processing of information.
- It helps in retrieval of information.
- It helps in performing arithmetic operations.
- A manager considers the electronic computer as a central element in an information system.

The three changes that occur in the management information system are :

- Management becomes more system oriented.
- Management becomes more sophisticated in using management techniques.
- Information is planned.
- Information is available to the managers whenever required.

Q.7. What steps are required for the evaluation of MIS? Explain.

Ans. The evaluation of the Management Information System is done after a small period of time. The evaluation is performed on each step of the :

- Design
- Final system performance

In the evaluation of a system, principle of feedback is applied. The process of evaluation should be performed as early as possible because if it is delayed, it becomes difficult for the designer to remember each and every important detail.

Key People : The key people involved in the evaluation of a management information system are :

1. Customers
2. Designers
3. Financial Specialists

The evaluation is performed by both the designers and customers. The financial specialists should evaluate the project in the following terms :

- Planned cost vs. actual cost of the design of the system
- Implementation
- Operation

Concepts of the Evaluation Process : The concepts that are involved in the evaluation process are :

1. **System Integrity :** The subsystems are integrated into the total system without any redundancy. It provides the following advantages :
   - Flexibility
   - Expandability
2. **Operation Integrity :** The operation integrity involves skilled people that operate the system. A backup should be available that should prevent the system in case of a system failure.
3. **Internal Integrity :** The internal integrity provides the following factors :
   - The validity of the output of the system.
   - Security of the system against human error, manipulation or theft.
4. **Procedural Integrity :** The procedural integrity involves :
   - Proper documentation of the system and procedures.

Structure of the Evaluation Process : The structure of the evaluation process is measured in terms of :

Measurement of cost or benefit = Measurement of Old – Measurement of New

The measurement of change is related to the basic goals of a management information system. It is the responsibility of the designer to select the level at which the measurement process should take place.

When the proper variables of interest are identified, a table should be set to formalize the measurement process.

Q.8. Write short notes on :

(*a*) Develop forms for data collection :

Ans. The input of the Management Information System is the large amount of data. The data can either be internal or external to a company.

For example : In a sales estimation, the salesperson must sit in a room and fill the form summarizing the day's activities.

The major benefit of data collection through forms is :

- They ensure that right information is provided.
- It helps in simplifying the data processing that is required for computer storage.
- Forms are also required for sending data at intermediate stages.

(*b*) Test the system.

Ans. The testing of the system is done on the basis of the test specifications and procedures. Testing during the installation system requires the following :

- Component tests
- Subsystem tests
- System acceptance tests

In the component testing, each component of the system is tested. In the subsystem testing, subsystems of the organization are tested. Acceptance testing is performed on the entire system.

Components are :

- **Equipment :** These can either be old or new.
- New forms
- New software programs
- New procedures

Performance meters : Tests are performed for the following meters :

- Accuracy of the system
- Range of inputs
- Frequency of inputs
- Reliability

Requirements of System Testing

- Verification of multiple inputs
- Verification of logic system
- Timing aspect of the parts of the system
- Interface of the system

Q.9. Explain management information system and the system approach.

Ans. The functions of management are classified into three parts :

- The capabilities of the organization should be matched with the needs of the organization.
- Establishment of a work environment.
- Allocation of resources so that they produce maximum productivity.
- Manage different responsibilities.

General Management System : The general management system consists of four factors :

(*a*) Planning : The planning consists of setting up of objectives, noticing opportunities, select the best course of action and design a program to achieve these objectives.

(*b*) Organizing : The organizing consists of supervising the action plan. The plan consists of the relationship between the tasks

(*c*) Directing : The directing consists of providing directions to the programs, communicating with the subordinates and motivating the subordinates.

(*d*) Controlling : The controlling consists of observing the performance, measuring the performance against the standards, observing the significant trends and modification of goals as per requirements.

(*e*) Feedback : The feedback consists of recycling of information, recycling of actions and ensure that the objectives are fulfilled.

Planning : Planning is the most basic management function. Planning consists of :

- What has to be done?
- When it has to be done?
- How it is to be done?

The planning helps us to reduce the gap between where we are and where we want to go.

Dimensions of Plan : The dimensions of plan can be classified as long range planning, medium range planning and short range planning.

(*a*) Long range planning : The features of long range planning are :

- Operates at corporate level
- The function is general management
- Companywide scope

(*b*) Medium range planning : The features of medium range planning are :
- Operates at division level
- The function is marketing
- Business scope

(*c*) Short range planning : The features of short range planning are :
- Operates at operational level
- The function is of scheduling the production process
- Business scope

Information and Planning : The information and planning is classified into three categories :

(*a*) Environmental Information : The environmental information can be further classified into four parts :

1. Political and Government : The features are :

(*i*) It is required for forecasting the plans.
(*ii*) It consists of financial policies.
(*iii*) It consists of tax policies.
(*iv*) It has an effect on planning decisions

2. Economic trends : The features are :

(*i*) Employment
(*ii*) Productivity
(*iii*) Capital investment
(*iv*) Price
(*v*) Wage levels

3. Technological Environment : The features are :

(*i*) Forecast the technological changes

4. Factors of production : The features are :

(*i*) Cost
(*ii*) Labor
(*iii*) Location
(*iv*) Availability

(*b*) Competitive Information : The competitive information can be further classified into three parts :

1. Industry demand : The features are

(*i*) Forecasting of demands by the firm

2. Firm demand : The firm demand is the function of the :

(*i*) Industry demand
(*ii*) Capabilities of the individual firm relative to the competing firm

3. Competition : The competition can further be classified into three parts :

(*a*) **Past Performance :** The features are :
- Profit
- Return on investment

- Market share
- Helps to identify competition

(b) **Present Performance :** The features are :

- Price
- Strategy
- Advertisement

(c) **Future Performance :** The features are :

- Information regarding new products
- Research and development

(c) Internal Information : The internal information can be further classified into four parts :

(i) **Sales forecast :** The features are :

- Important planning document

(ii) **Financial Plan :** The features are :

- Involves the entire organization
- Provides information for sub-plans

(iii) **Supply factors :** The features are :

- Labor
- Capital
- Organization
- Equipment

(iv) **Policies :** The features are :

- Product
- Marketing
- Financial policies
- Term policies

□□□

PREVIOUS YEAR QUESTION PAPER-4

Q.1. Attempt all questions in brief.

(*a*) Explain file design.

Ans. The file design consists of the logic that is required to control the flow of data through the system. It is a part of the system design only. It should be noted that the flow of data depends solely on the design of the data files.

(*b*) What are the five types of internal constraints?

Ans. Top management support, policy considerations, personal needs, cost and restrictions are the various types of internal constraints.

(*c*) Explain restriction internal constraint.

Ans. Restriction is the constraint that is made by the managers and designers on the design of an information system. The manager needs to alter some of the objectives so that the system is fit with other outputs and equipment. The manager will also restrict on the time and effort that is applied to the entire phase.

(*d*) What are the components of operational control?

Ans. The operational control includes formal reports, fixed procedures and daily transaction report.

(*e*) Explain external records.

Ans. External record is the record that comes from various different sources. Example is government statistics.

(*f*) Explain parallel approach.

Ans. In the parallel approach, the selected new system is installed and operated with the current system in parallel. It is an expensive approach.

(*g*) What are the key people involved in the evaluation process?

Ans. The key people involved in the evaluation of a management information system are customers, designers and financial specialists.

(*h*) Explain the creation era.

Ans. The word supply chain management was invented by the U.S. industry consultant in 1980's. The characteristics of the creation era are large scale changes, re-engineering and cost reduction programs.

Q.2. Explain the computer software in detail.

Ans. The computer software is a collection of computer programs. In contrast to the computer hardware, the software cannot be touched. These programs provide instructions to the computer for what and how to do. Thus, software is a set of :

- Programs
- Procedures
- Algorithms

The functions are performed either directly or indirectly. It is done directly by providing instructions to the computer hardware. It is performed indirectly by serving as in input to other software.

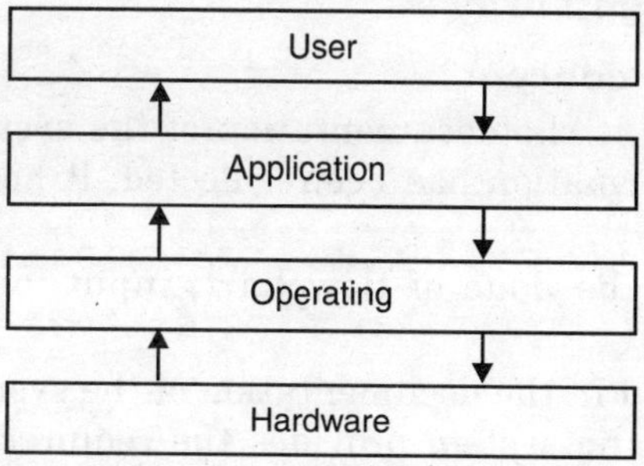

System Software : System software is responsible for controlling, integrating, managing the individual hardware components of a computer system. The users are not concerned with the low level details like transferring of data from the memory to disk.

Components of System Software : The components of system software are :

- Device drivers
- Operating systems
- Servers
- Window systems
- Management tools
- Networking control software
- Disk formatters
- File managers

Functions of system software : The functions performed by the system software are :

- It is responsible for managing independent hardware components.
- Help the application software programmer from the complex details of the system.
- Partition the computer resources in a safe and stable manner.

For Example : Microsoft word is developed in languages like C, C++.

Types of Software : The classification of software is done in two parts :

(*a*) **Programming Software :** The programming software consists of tools in the form of programs and applications. The software developer makes use of the programming software to perform the following functions like create the programs, debug the programs, maintain the programs and support other programs.

For example : The examples of programming software are compliers, debuggers, interpreters, linkers and text editors.

(*b*) **Application Software :** The application software is basically a set of programs that performs specific data processing. It consists of many softwares.

Computer Software Architecture : The computer software architecture consists of three layers. These are platform software, application software and user written software.

(*a*) **Platform Software :** The platform software allows the user to interact with the computer and its peripheral devices. It consists of firmware, device drivers, operating system and graphical user interface.

(*b*) **Application Software :** Application software is the independent programs that are modified according to the platforms. They are purchased separately from the computer hardware. **For Example :** Microsoft office suite and video games.

(*c*) **User-written software :** User written software is the software that is created by the user. **For Example :** Spreadsheet templates and word processor templates.

Q.3. Explain the following in brief :

(a) Document the system concept :

Ans. Now the major factor is the documentation of the system concept. Documentation is required because large information has been collected. It helps in making a more detailed description of the system.

Documentation should be done of flowchart, input, output, operations and flow of information.

Manager is also involved in the documentation of the system. The manager is concerned with the amount to which the system provides the required information. Designer is also involved in the documentation of the system. The designer is related to materials, equipment and technical processes.

Black Box Concept : The black box concept includes the following elements :

- Input
- Output
- Master files
- Data processing rules

There are certain rules that should be followed in the black box concept. These are :

- Processing of the input data against the data file and finally produce an output.
- Processing the input data for updating the file.
- Processing of input data to output data without referencing the data file.
- Producing an output from the file without processing the input.

Generate System Flow : The generation of system flow is a method of specifying the general structure of a computer based information system. It reflects the design effort.

Characteristics : The characteristics of the system flow are :

- System objectives are easily achieved.
- The system objectives are reflected in the flow diagram.
- Information needs are designed into the system.
- Information sources are designed into the system.
- Subsystems are integrated into one system.
- Decision rules and points are shown.

System Inputs : The system inputs are the ones that are received from outside the firm and are converted to machine usable form. The documentation of system inputs must be done in a detailed manner. The documentation must include the following :

- Source of the input data
- What is the form of the input data?
- Who is responsible for the production of the input data?

System designers are concerned with the system inputs. System inputs are collected with the help of forms. Thus it is the responsibility of the designer to design the form in an appropriate manner.

System Outputs : The system output is the one that is received from the processing of the input data. The documentation must include the following :

- Where does the system output go?
- What form they take?
- Who is responsible for receiving them?
- How many copies of the output form are required?

• What kind of output form is required?

Documentation of Inventory Management : The documentation of the inventory management of the raw materials is concerned with the raw materials that are required for the manufacturing process. The inventory management of raw materials includes four plants. Each plant has its own computers. The three subsystems present are :

- Stock level subsystem
- Reorder point subsystem
- Economic order quantity

Documentation of Stock Level : In the documentation of stock level, as soon as the material is received, the receiving report is marked to show :

- Quantity acceptance
- Quality acceptance

There is a matching done between the received report and receiving order about the :

- Quantity acceptance
- Quality acceptance
- Data acceptance

Finally the receiving order is sent to the computer for updating the inventory file.

Q.3. (*b*) Prepare Conceptual Design Report.

Ans. A conceptual design report is prepared to know the amount spend in the organizational changes. The design report must consist of the following parameters :

- Objectives
- Summary
- Nature of the system
- Reason of selecting alternative design
- Time required to design the system
- Time required to implement the system
- Resources required to design the system
- Resource required to implement the system

Q.4. Explain the following :

(*a*) Input, output and processing.

Ans. Input and Output : FORMS : The **input** is that each subsystem requires information. **Output** is the information that is produced by processing the input so that it can be used by other subsystems. The input and output operations are normally done with the help of **forms.** Form is a communication link that is used in the operating procedure.

The major tasks in performing this operation are :

- Input should be specified properly
- Output should be specified properly
- Design the forms that are relevant
- Design the forms that are easy to use

Identification Task

- Functions of the form
- No of copies required
- Where is the form to be used

- Who will use the form
- How many units of form are required per year

Form Designing Factors : The following factors should be considered while designing the forms :

- Forms should be attractive
- Forms should be easy to read and use
- Forms should contain an identification number
- Forms should contain the date of issue
- Forms should be designed in a logical manner either from left to right or top to bottom
- Forms should contain sufficient data to enter the data

Input and Output : DATABASE : Database is a data that can be stored and retrieved. It helps in making decisions. The decisions can either be managerial decisions or programmed decisions.

Approach : The approach that must be followed in the development of database is :

- Identification of all the points in the flowchart that require input data.
- Prepare data for each data element providing the :
  - Source of data
  - Length of data
  - Format of data
  - End use of data
- Group all the data with activity
- Check for errors
- Check if any duplicate data is present
- Eliminate the duplicate data so that an integrated database is developed

Processing : In the processing following factors are there :

- Vast amount of information has been gathered
- Input is present in a detailed manner
- Output is present in a detailed manner
- Corrections should be made to the subsystem
- The task that is specified in the subsystem should be broken down either in a mathematical manner or narrative format.

(*b*) Software, hardware and tools

Ans. The software, hardware and tools consist of steps that must be followed. It is the responsibility of the system designer to check these steps in a proper manner.

Steps for software : The steps for software are :

- Develop standards for programming.
- Develop programs for programming.
- Develop the logic.
- Develop programming flowcharts.
- Study the conceptual system design specifications.
- Work with the system designer in a team to develop detailed design.

Steps for hardware : The steps for hardware are :

- Enough computing power must be purchased to perform the whole job.
- Enough external storage must be purchased to store the required database.

- Peripheral devices must be purchased to support the MIS in solving the problems.
- Computer supplies must be purchase. For example : paper, tapes etc.
- Best documentation must be purchased.
- Maintenance package must be purchased.

(c) Propose an organization to operate the system

Ans. An organization must be proposed to operate the system. The management information system group should work with the top managers to suggest changes that will fulfill the requirements of the new system. It is the responsibility of the subsystem managers that the subsystem should work in the best possible way. They should also try to increase the performance of the system.

Q.5. Explain the following :

(a) Computer related acquisitions.

Ans. The computer related acquisitions can either be hardware or software.

(a) Hardware :

- In computer hardware, the design of the computer architecture available from the vendors and the system are closely tied together.
- A choice of CPU and peripheral devices are to be made.
- In this, it has to be decided whether the hardware is to be purchased or to be brought on lease.
- Capital analysis, usage etc. are the factors that are to be taken into consideration.

(b) Software :

- Many software packages are available commercially.
- Small companies must purchase the software products. But, the purchasing of software has some problems. Sometimes the software requires so many modifications that it would be cheaper to develop the software internally.
- Large companies can either develop the software internally or under contract.

(c) Personnel :

- It consists of a personnel planning chart. This chart shows the number of persons required in terms of skill, source etc. The source can either be internal or external.
- It provides an opportunity to upgrade the personnel after the training process.
- At higher levels, computers must be used to handle the jobs which were earlier performed by individuals manually.

(d) Materials :

- The two main materials that are required in a Management Information System are forms and manuals.
- Demand for the forms and manuals must be estimated so that it can be ordered.
- Computer materials like cards, tapes, printout paper should be ordered whenever required.

(b) Develop the files.

Ans. The actual data must be obtained and recorded for the initial testing and operation of the system.

The requirements of developing these files are :

- Format of data
- Storage form

- Storage format
- Remarks to show when the data has been stored

Procedures are required to be developed for updating each piece of data and the entire file.

(*a*) **Mater file :** Master file is a file that collects the data that is used in various operations. A procedure must be developed to obtain the data from the environment. The responsibility of maintaining each file must be assigned to each person. The designing of forms should also be done. The form helps the programmer to analyze the data.

Requirement for each translation :

- File name
- Volume of operations
- Maximum number of characters required to each record element

Q.6. Explain evaluation and control of MIS.

Ans. The evaluation of the Management Information System is done after a small period of time. The evaluation is performed on each step of the design and final system performance.

In the evaluation of a system, principle of feedback is applied. The process of evaluation should be performed as early as possible because if it is delayed, it becomes difficult for the designer to remember each and every important detail.

Key People : The key people involved in the evaluation of a management information system are customers, designers and financial specialists. The evaluation is performed by both the designers and customers.

Concepts of the Evaluation Process : The concepts that are involved in the evaluation process are :

1. **System Integrity :** The subsystems are integrated into the total system without any redundancy. It provides the following advantages :
   - Flexibility
   - Expandability
2. **Operation Integrity :** The operation integrity involves skilled people that operate the system. A backup should be available that should prevent the system in case of a system failure.
3. **Internal Integrity :** The internal integrity provides the following factors :
   - The validity of the output of the system.
   - Security of the system against human error, manipulation or theft.
4. **Procedural Integrity :** The procedural integrity involves :
   - Proper documentation of the system and procedures.

Structure of the Evaluation Process : The structure of the evaluation process is measured in terms of :

Measurement of cost or benefit = Measurement of Old – Measurement of New

The measurement of change is related to the basic goals of a management information system. It is the responsibility of the designer to select the level at which the measurement process should take place.

When the proper variables of interest are identified, a table should be set to formalize the measurement process.

Control the Management Information System : The control of the Management Information System is done by the line managers. The control of the system means that the operations should operate as it were designed to operate. The operators must develop their own procedures or may modify the procedures that were already designed.

Q.7. Write short notes on :

(*a*) System Planning.

Ans. A proper system planning is required to perform the maintenance of the management information system in use. Maintenance cannot be performed on a first come first serve basis.

The factors that must be taken into account are :

- Record all the requests that are made for the change process. Only requests that are made in a written format should be accepted and recorded.
- A priority should be given to all the requests. The priority is made on the following factors :
  - Urgency of the project
  - Long term benefits provided by the project
  - Time required
  - Resources required
- Make annual and monthly plans.
- Document the maintenance process.

(*b*) Implementation problems.

Ans. Various types of implementation problems are faced in the development of a management information system.

Review

- **Hardware Perspective :** In the hardware perspective, the managers show off their computer centers, disk space and the millions of instructions the processor executes per second. Nowadays the hardware items are becoming cheap and are of little concern in the MIS purchase design.
- **Software Development :** The software development pitfall includes two pitfalls. These are hardware related and personnel related. The hardware performs the instructions. Therefore wrong instructions lead to wrong results.
- **Test it and Test It Again :** A major error that is seen is improper planning that is done to perform the testing. Testing must be performed at the functional level, component level and system level. Problems are experienced at each level and they must be fixed.
- **Controlling the MIS Project :** Controlling the MIS Project is the one of the four basic management functions. It also remains one of the major causes of MIS development project failure.

Q.8. How is MIS maintained? Explain.

Ans. The maintenance of the management information system is very similar to the controlling of the system. Maintenance is basically an activity that helps to keep the MIS system at the highest level of effectiveness and efficiency within the cost limitations. It helps in reducing the errors that are caused by the designers and environmental change.

The maintenance process is applied to the following activities :

- Change in the forms
- Change in the operating system
- Change in the procedures that are used
- Modification in the hardware
- Modification in the software
- Changes in the policy statements

The maintenance activities are classified into three types. These are :

1. Environmental Change : The environmental change includes four factors. These are described below :

(*a*) Governmental Policies : Specialists are required in many companies. The role of these specialists is to remain updated about the various changes that take place in the requirement process. The governmental policies, regulations and legislations keep on changing from the government side. These factors require constant updating of the MIS in a company.

(*b*) Economic Conditions : The economic conditions play an important role in defining the financial information of the system. The system should be properly defined. It should meet the needs of the users, accounting department and finance department.

(*c*) Industrial Conditions : The industrial conditions are quite similar to the economic conditions. The main difference between these two conditions is that the industrial conditions reaction to changes is very important. The areas that are affected by the industrial conditions are price, policy, hiring methods and budget.

(*d*) Technological Conditions : New technology is created and is used by a large number of people. Continuous system maintenance is required for this factor.

Internal Problems in System Maintenance : There are many problems that are related to the maintenance of the management information system. These are :

- No proper plan
- No proper allocation of resources
- Lack of management understanding problems
- Improper documentation
- Lack of understanding

System Maintenance Responsibility : The maintenance is done by specific people. These are :

- Specific responsibility is given to the supervisor.
- Special team of MIS analyst, specialist and programmer is formed to perform system maintenance.

If the tasks are not assigned to special people, problems may arise.

Maintenance Projects : Maintenance projects are created to form :

- Error reports
- Request changes by the users
- Member of the maintenance team
- Management of the company

Q.9. Write short notes on :

(*a*) SCM.

Ans. Supply chain management are a set of approaches that are used to efficiently integrate the suppliers, manufacturers, warehouses and distribution centres. It is also the management of a network of interconnected businesses that are involved in the delivery of product and service packages that is required by the end customers. Supply chain management is also known as logistics network. Supply chain management is the designing, planning, controlling, monitoring of the supply chain activities with an objective of creating net value, measuring performance globally and synchronizing supply with demand.

There are three inputs of supply chain management. These are raw material, working ininventories and finished products.

(b) Business process integration.

Ans. Supply chain business process integration is the collaboration between buyers and suppliers. According to Lambert and Cooper, integrated supply chain requires a continuous information flow. The key supply chain processes stated by Lambert are :

- Customer relationship management
- Customer service management
- Demand management style
- Order fulfilment
- Manufacturing flow management
- Supplier relationship management
- Return management
- Product development
- Product commercialization
- Other key critical supply chain business processes started by Lambert are :
- Procurement
- Performance measurement
- Warehousing management
- Manufacturing flow management
- Physical distribution

(c) Procurement Management System.

Ans. A procurement management process or a procurement process is the method of purchasing goods and services from external suppliers. Obtaining goods and services from external suppliers can be a critical path for many projects. The performance of the supplier reflects the performance of the overall project team. The supplier's performance should be monitored carefully and regularly so that the delivery of the product meets the expectations of the team.The goods and services should be obtained at the best possible cost to meet the needs of the purchaser in terms of quality, quantity, time and location. The management process provides procurement procedures. This procedure consists of step-by-step explanation of how to purchase goods from suppliers

Thus the procurement management process helps you : "Get What You Have Paid For".

(d) Reasons for ERP implementation.

Ans. Enterprise resource planning is an integrated computer based application that is used to manage internal and external resources, financial resources, materials and human resources. Some of the top reasons for implementing ERP are :

1. **Integration of systems** : Integrated business systems help the department to work more efficiently and effectively. It helps the organization to improve production cycles. It helps in managing the work processes and gets an easy view of the work flow.
2. **Enables new business to grow :** It helps a growing organization to handle number of tasks efficiently through ERP software. Understanding of the complete work flow enables the organization to remove the loopholes in functioning.
3. Internet and wireless networks helps the organization to grow and publicize their products. They also help to stay in touch with customers and business partners.
4. **Reduce expense on business processes :** The enterprise system helps the organization to work efficiently which helps in reducing the expenses on business processes.

5. **Effective decision making** : The management can take decisions with the help of previous data of the company and online services.
6. **Technological advancement for extending business services** : The business services can be extended with the help of technological advancements. Internet plays an important role in this. With the help of Internet, business processes can be promoted online rather than spending large amount of money by marketing manually.
7. **Improve communication with partners and customers :** ERP system enables an organization to send emails and messages to the customer without manual intervention. It also helps to provide easy customer service. The ERP software helps the company to retain the client and increase the client base.

❑❑❑

Index

Win Prizes

ATTENTION READERS

Win Prizes

We request for your frank assessment regarding various aspects of the book as given below :

MANAGEMENT INFORMATION SYSTEM

By: NEHA GHAI

Please Fill up the Sheet in Neat and Clean Writing

(i) What topics of your syllabus, important for your examination point of view, are not covered in this book ?

....

....

....

....

(ii) What are the chapters wherein the treatment of the subject matter is not systematic or organised or updated ?

....

....

....

....

(iii) Have you come across misprints/mistakes/factual inaccuracies in the book? Please specify the chapters and the page numbers.

....

....

....

....

(iv) What other books on the same subject have you found/heard better than the present book? Please specify in terms of price and quality

....

....

....

....

(v) Further suggestions and comments for the improvement of this book.

....

....

....

....

(PS: If need be please attach a separate sheet to write your views)

OTHER DETAILS

(i) Who recommended you this book ? (Please tick (✓) in the box near the option relevant to you.

☐ Teacher ☐ Friends ☐ Bookseller

....

....

....

....

(ii) Name of the recommending teacher, his designation and address :

....

....

....

....

(iii) Name and address of the bookseller you purchased the book from :

....

....

....

....

(iv) Name and address of your institution (Please mention the University or Board as the case may be)

....

....

....

....

(v) Your name and complete postal address :

....

....

....

....

(vi) Course you are enrolled in :

....

....

....

....

The best assessment will be awarded each month: The award will be in the form of our publications as decided by the Editorial Board.

Please mail the filled up coupon at your earliest to :

S.K. KATARIA & SONS®
4760-61/23, Ansari Road,
Daryaganj, New Delhi-110002
☎: +91-11-23243489, 23269384
Email : katariabooks@yahoo.com
katariabook@gmail.com